28th International Carbohydrate Symposium (ICS 2016)

New Orleans, Louisiana, USA
17 - 21 July 2016

ISBN: 978-1-5108-3585-6

Printed from e-media with permission by:

Curran Associates, Inc.
57 Morehouse Lane
Red Hook, NY 12571

Some format issues inherent in the e-media version may also appear in this print version.

Printed by Curran Associates, Inc. (2017)

International Carbohydrate Symposium (ICS)
1155 16th St NW
Washington, DC 20036
USA

Phone: (202) 452-2138

ics2016@acs.org

Additional copies of this publication are available from:

Curran Associates, Inc.
57 Morehouse Lane
Red Hook, NY 12571 USA
Phone: 845-758-0400
Fax: 845-758-2633
Email: curran@proceedings.com
Web: www.proceedings.com

28th International Carbohydrate Symposium

I.C.S.2016

XXVIII International Carbohydrate Symposium

July 17-21, 2016 • New Orleans, Louisiana, USA

TABLE OF CONTENTS

I.C.S.2016

XXVIII International Carbohydrate Symposium

July 17-21, 2016 • New Orleans, Louisiana, USA

Dear Friends and Colleagues,

On behalf of the International Carbohydrate Organization (ICO), the U.S. Advisory Committee for International Carbohydrate Symposia (USACICS) and the Organizing Committee, I welcome you to New Orleans for the 28th International Carbohydrate Symposium.

The biennial International Carbohydrate Symposia (ICS) are considered to be the most important meetings for carbohydrate science. The ICS are the main activity of the ICO, which in turn is an Associated Organization of the International Union of Pure and Applied Chemistry. The USACICS is the organization designated by the American Chemical Society (ACS) to interact with the ICO, and maintains funds for the Whistler award and other promotions of carbohydrate science. The USACICS is has representatives from the ACS Division of Carbohydrate Chemistry and the Division of Cellulose and Renewable Materials. ACS Conference Management is providing our infrastructure, so they are visible here, as well. The organizers have received very substantial help in putting the meeting together from the International Advisory Committee, the Scientific Committee, and the Local Committee. In particular, the technical program was organized by members of the Scientific and Organizing Committees, augmented by Rob Woods, Maren Roman, Robert DiCosimo, Nicole Labbé, and Waldemar Priebe.

The scientific program covers all aspects of modern basic and applied glycoscience, with eight plenary lectures, 32 invited keynote lectures, 195 contributed oral presentations, and about 200 posters. Additionally, the R. L. Whistler Award and the the ICO Young Researcher Award will be presented by the ICO, and the C. S. Hudson Award will be presented by the Carbohydrate Division of the American Chemical Society. Our congratulations to the winners, Benjamin Davis (Oxford University), Geert-Jan Boons (University of Georgia), and Benjamin Swarts (Central Michigan University), respectively. All will present award addresses.

We are very grateful for our generous sponsors as well as to all who have come from around the world to contribute to this meeting. Our optional tour program is one of the most ambitious for an ICS and the social program is especially reflective of the local culture which has made New Orleans one of the top tourist destinations as well as a very popular convention city. We hope that you have a great meeting.

Sincerely,

Alfred D French

Alfred French
Symposium Chair

MITCHELL J. LANDRIEU, MAYOR

CITY OF NEW ORLEANS

Welcome!

As Mayor of New Orleans, it is my pleasure to welcome participants of the 2016 International Carbohydrate Symposium. Thank you for your commitment to host your symposium in July.

You will experience the warm hospitality and unique culture that New Orleans has to offer. I am certain that New Orleans will serve as the perfect destination for this gathering and provide the best opportunities for the group. While you are here, take some time to relax and explore the City. Dine in our incomparable restaurants; enjoy some of the great music we have to offer, tour historic neighborhoods like the French Quarter and Treme, and shop in our antique and specialty shops throughout New Orleans, including the bustling Magazine Street corridor. It is my hope that your stay will be memorable and that you leave with new knowledge, relationships and a uniquely unforgettable experience.

I extend my best wishes to all symposium participants and those traveling to partake in the event. Thank you for choosing New Orleans.

Sincerely,

Mitchell J. Landrieu, Mayor
City of New Orleans

1300 PERDIDO STREET | SUITE 2E04 | NEW ORLEANS, LOUISIANA | 70112
PHONE 504-658-4900|FAX 504-558-4938

UNITED STATES OF AMERICA

State of Louisiana

John Bel Edwards
GOVERNOR

Proclamation

WHEREAS, the International Carbohydrate Symposium is a conference of the most prominent carbohydrate research scientists from around the world, sanctioned by The International Carbohydrate Organization and sponsored by the International Union of Pure and Applied Chemistry; and

WHEREAS, this conference, which rotates between Europe, North America, Asia, and Australia, which last met in the United States in 1998, and will be meeting in Louisiana for the first time in July 2016; and

WHEREAS, it is appropriate for New Orleans and Louisiana to host this symposium as there is significant research on carbohydrates in Louisiana institutions and two very significant agricultural carbohydrates, cotton cellulose and sugar, which both have long histories in the economy of Louisiana; and

WHEREAS, carbohydrates have been recognized as being instrumental in cell-to-cell surface recognition and interactions, in immunology and other interactions that can lead to better understanding how to fight viruses, bacteria, cancers and other human and animal diseases; and

WHEREAS, the latest research on carbohydrates, including food, fiber, energy (biofuels), and medical fields such as human physiology, disease recognition, and carbohydrate-based vaccines will be reported at the Symposium with Plenary, Invited and Contributed Lectures as well as Posters; and

WHEREAS, the Roy L. Whistler Award, the most important recognition of a carbohydrate scientist in the world, will be presented at the Symposium.

NOW, THEREFORE, **I, John Bel Edwards**, Governor of the State of Louisiana, do hereby proclaim July 17-21, 2016 as

INTERNATIONAL CARBOHYDRATE WEEK
in the State of Louisiana.

In Witness Whereof, I have hereunto set my hand officially and caused to be affixed the Great Seal of the State of Louisiana, at the Capitol, in the City of Baton Rouge, on this 10th *day of* May *A.D.,* 2016 .

Governor of Louisiana

ACS
Chemistry for Life®

ACS Conference Management

ACS manages hundreds of meetings a year from local gatherings to nationally- and internationally-renowned exhibits and conferences drawing scientists from around the world.

Our Services

Site selection and contract negotiation – Recommend and research venues and negotiate a contract that best meets your needs in terms of room rates, meeting space and food and beverage requirements

Housing – Manage housing for your conference, set up online room reservations, room block management and online reports

Registration – Set up online registration, collect revenue, provide badges, and on-site meeting management

Technology – Provide 24/7 access to registration and meeting information

Logistics – Coordinate audio-visual, room set-up, and food and beverage needs

Financial Services – Develop and manage budget, collect revenue, pay bills and report activity

Marketing – Coordinate advertising in publications and production of programs

Abstract Management – Provide an abstract management system to collect abstracts, organize presentations and provide a printed program, abstract books and CDs.

Reporting – Pre- and post-meeting reports on finances and attendance

Package Plans – Every group is different. ACS recognizes that you many have a pre-established relationship with some trusted contractors and vendors.

NOTES

28th INTERNATIONAL CARBOHDYRATE SYMPOSIUM

Advion

Alnylam PHARMACEUTICALS

BRUKER

Carbosynth

DIVISION OF CARBOHYDRATE CHEMISTRY

Cellulose and Renewable Materials Division
A Division of the American Chemical Society

The University of Georgia Complex Carbohydrate Research Center

CORDENPHARMA

Dextra

epnoe European Polysaccharide Network Of Excellence

ELSEVIER

IUPAC International Union of Pure and Applied Chemistry

J&K Scientific

LC scientific inc.

Louisiana Section ~ New Orleans ~ 1906

Megazyme

Mizutani Foundation for Glycoscience

Omicron Biochemicals, Inc. .....embracing sugars for science and society

Organic & Biomolecular Chemistry

PROzyme Advancing Glycosciences

USDA

Sussex Research

TCI TOKYO CHEMICAL INDUSTRY

V-LABS INC.

WYATT TECHNOLOGY

YAMAZEN CORPORATION Japan's best-selling Automated Flash Systems

About the International Carbohydrate Organization

The International Carbohydrate Organization (ICO) was formally established in 1980. Prior to that, Carbohydrate Symposia were coordinated by an International Steering Committee that first formed in 1965. The ICO comprises representatives from some 33 countries accredited by the ICO. The aims of the ICO are to encourage research, communication and education in the glycosciences.

The ICO has been an Associated Organization of the International Union of Pure and Applied Chemistry (IUPAC) since 1970 and an Interest Group of the International Union of Biochemistry (and Molecular Biology) since 1982.

With the rapid advances in carbohydrate chemistry and increasing interest in carbohydrate meetings a new series of meetings, restricted to Europe, started under the name of Eurocarb in Vienna in 1981. These meetings take place biennially in those years in which there is no International Carbohydrate Symposium. Eurocarb is modelled on the International Symposia and is controlled by the European Carbohydrate Organization, representing countries in Europe in which carbohydrate research is carried out.

The International Carbohydrate Symposium (ICS) is held every two years in a selected host country on behalf of the ICO. The ICS is the premier meeting venue for scientists working in all fields of carbohydrate science.

Future International Carbohydrate Symposia (ICS) are planned in the following countries:

2018 Portugal

2020 Brazil

2022 China

2024 Slovakia

2026 Canada

2028 Italy

2030 Taiwan

President
N. Jayaraman, India

Secretary
Spencer J. Williams, Australia

Past President
Jesús Jiménez-Barbero, Spain

President-Elect
Zbigniew J. Witczak, United States

"The International Carbohydrate Symposium is the most important event in carbohydrate chemistry and biochemistry."

Organizing Committee, 25th ICS, Tokyo

National ICO Representatives

ARGENTINA
Oscar Varela

AUSTRALIA
Spencer J. Williams

AUSTRIA
Paul Kosma

BRAZIL
Philip A.J. Gorin

CANADA
Todd Lowary

CHINA
Biao Yu

CUBA
Vincente Verez Bencomo

CZECH REPUBLIC
Vladimir Kren

DENMARK
Mikael Bols

FINLAND
Risto Renkonen

FRANCE
Pierre Monsan

GERMANY
Thisbe K. Lindhorst

HUNGARY
Lajos Szente

INDIA
N. Jayaraman

IRELAND
Paul Murphy

ISRAEL
Timor Baasov

ITALY
Francesco Nicotra

JAPAN
Yukishige Ito

KOREA
Injae Shin

THE NETHERLANDS
Johannis P. Kamerling

NEW ZEALAND
Richard Furneaux

NORWAY
Berit Smestad Paulsen

POLAND
Slawomir Jarosz

PORTUGAL
Amelia P. Rauter

RUSSIA
Yuriy Knirel

SLOVAKIA
Miloš Hricovinia

SOUTH AFRICA
David W. Gammon

SPAIN
Jesús Jiménez-Barbero

SWEDEN
Olof Ramström

SWITZERLAND
Beat Ernst

TAIWAN
Chun-Hung Lin

UNITED KINGDOM
Rob Field

UNITED STATES
Zbigniew J. Witczak

COMMITTEES

Organizing Committee

Alfred French – Chair
Zbigniew J. Witczak – Vice Chair
Sheila Murphy – Program Chair
Mary An Godshall – Executive Secretary

International Advisory Committee

Yuriy Knirel
N. Jayaraman, ICO President
Timor Baasov
Jesús Jimenez Barbero
Antonio Molinaro
Yukishige Ito
Paul Kosma
Berit Smested Paulsen
Otto Holst
Brian Mario Pinto

Scientific Committee

Sheila Murphy – Program Chair
Geert-Jan Boons
Mary Cloninger
Alexei Demchenko
Laura Kiessling
Kelley W Moremen
James Paulson

U.S. Advisory Committee for ICS

Zbigniew Witczak
Jim BeMiller
Peter Andreana
Alfred French
Lai-Xi Wang
Sheila Murphy
Alexei Demchenko
Anthony Serianni
Nicole R. Brown
Nicole Labbé

Local Committee

H.N. Cheng
Xavier Miranda
Marie Alice Rousselle
John Vercellotti
Sharon Vercellotti

OUR SPONSORS AND EXHIBITORS

ACS Division of Carbohydrate Division The central mission of the Division of Carbohydrate Chemistry is the stimulation of interest, the encouragement of research, and the dissemination of information in all that pertains to the chemistry and technology of carbohydrates. Currently, there are around 750 domestic and international division members from academia, industry as well as the government, encompassing research areas including carbohydrate chemistry, glyco-biochemistry, glycobiology, glycomics, glycomimetics, glyco-materials, glyco-nanotechnology, and glyco-engineering. The Division programs every year at the American Chemical Society annual national meetings and actively sponsors other international, national and regional meetings on glycoscience. In addition, the Division sponsors several major international awards to researchers with outstanding contributions to the field.

ACS Division of Cellulose and Renewable Materials inspires bio-based solutions for a sustainable future. Its mission is to lead and support innovations in cellulose and renewable materials by providing a forum for our members to excel in the chemical sciences and technology. The Division of Cellulose & Renewalbe Materials currently has over 2,500 national and international members.

ACS Louisiana Section consists of 500-600 members in the Greater New Orleans area. We work variously in academia, industry, and government labs. We work hard but we also enjoy the beautiful music and the delectable food around here. Our local section is very active. We organize events for our members to interact and network, provide opportunities for recognition and professional development of our members, encourage chemistry education, and enhance the public's understanding of chemistry.

Advion, Inc. was founded in 1993 based on the novel techniques developed within the Cornell University laboratory of Dr. Jack Henion, a leading researcher in the field of Liquid Chromatography/Mass Spectrometry (LC/MS). The company's vision was to become a leading provider of bioanalytical contract services to the pharmaceutical industry, a position it attained prior to the sale of its contract drug development laboratory business to Quintiles. Pioneers within the industry, Advion offers a variety of mission critical systems and consumables for life science research. With sales and support offices in North America, Europe, and Japan, Advion is a global company with customers in all of the top pharmaceutical companies, government life science research agencies and universities. Advion continues to expand its diverse portfolio of innovative microfluidic and mass spec – based products for the life science industry.

Alnylam Pharmaceuticals was founded in June 2002 with the focus of advancing RNAi therapeutics as a new class of innovative medicines. Alnylam scientific founders developed a new strategy to trigger RNAi in mammalian cells using relatively small dsRNAs – long enough to induce RNAi, but small enough to avoid inducing an immune response. Alnylam founders were the first to show that smaller dsRNAs, known as "small interfering RNAs" (siRNAs) bind to messenger RNAs (mRNA) and silence disease causing genes. These discoveries opened the door for application of RNAI as a new therapeutic strategy.

Bruker designs and manufactures analytical instrumentation for elemental analysis, materials research, structural and surface investigations. Bruker's range of technologies contains XRD, SAXS, SC-XRD, X-ray Micro-CT, XRF, Micro-XRF, TXRF, FT-IR, NMR, and mass spectrometry. The Bruker portfolio also includes compact, powerful, robust, yet east-to-use instrumentation specially developed for scientific education purposes. These instruments are ideal for use by both students and experts in an academic setting. Bruker is strongly committed to fully meet customers' needs and continue to develop innovative solutions for today's analytical questions

Carbosynth Limited was established to provide the most comprehensive product range in the fields of carbohydrates and nucleosides for the international Life Science R&D communities. Carbosynth aims to support its clients from discovery to launch by offering cost effective solutions to their carbohydrate and nucleoside requirements. We produce many products in our own laboratories and distribute for CMS Chemicals Limited.

CCRC The Complex Carbohydrate Research Center, located on the campus of the University of Georgia, occupies a 140,000 sq. ft. building, specifically designed for the interdisciplinary and equipment-intensive nature of carbohydrate science. The CCRC is organized to optimize cooperation and collaboration among disciplines (biomedical, plant, and microbial glycosciences, synthetic and analytical chemistry) both within the CCRC and with scientists world-wide and to foster analytical service and training.

CordenPharma is your full-service partner in the Contract Development & Manufacturing (CDMO) of APIs, Drug Products and associated Packaging Services. Through a network of fully-inspected cGMP facilities across Europe and the US organized under five technology platforms – Peptides, Lipids & Carbohydrates, Injectables, Highly Potent & Oncology, Small Molecules, Antibiotics - CordenPharma experts translate complex ideas at any stage of development into high-value products.

Dextra Laboratories was founded in 1989 by two scientists who formerly worked at Tate & Lyle, with the objective of providing carbohydrate synthesis and analytical services. Since then Dextra Laboratories has gained a world-wide reputation for the process scale supply and manufacture of rare sugars and oligosaccharides for medical research, diagnostics and pharmaceutical applications. Dextra has built up an impressive 'toolbox' of techniques, strategies and building blocks enabling us to manufacture and analyse complex carbohydrates. At Dextra, many novel processes have been designed and undertaken leading to the successful synthesis of complex carbohydrates and API's including oligosaccharides, nucleosides, unnatural L-sugars, sulfated sugars, azasugars and natural products. To strengthen its service offering, a cGMP suite comprising two separate class 100,000 laboratories was installed. The laboratories accommodate state of the art equipment, such as 50L jacketed vessels, 5L pressure vessel and a range of analytical equipment, allowing the production of up to 5 10 kg of APIs suitable for phase I and phase II clinical trials. A formal quality management system compliant with cGMP requirements (Rules and Guidance for Pharmaceutical Manufacturers and Distributors, section 19) was established which enhances the day-to-day operation of the business and demonstrates Dextra's commitment to quality. The quality management system which covers all aspects of the business was audited by the British Standards Institution (BSI) leading to accreditation to ISO 9001:2008 in April 2009. In addition Dextra is regularly audited by our clients (typically US and European) which are often conducted to FDA/MHRA standards. Dextra prides itself in being a one-stop shop for Carbohydrate expertise.

Elsevier is a global business, headquartered in Amsterdam, The Netherlands, with offices worldwide. Elsevier is a world-leading provider of information solutions that help you make better decisions, deliver better care, and sometimes make groundbreaking discoveries in science, health, and technology. We provide web-based, digital solutions, among them – Science Direct, Scopus, Evolve, Knovel, Reaxys, and ClincalKey – and publish over 2,500 journals and more than 33,000 book titles. We publish Carbohydrate Research, Carbohydrate Polymers, and the Advances in Carbohydrate Chemistry and Biochemistry book series.

EPNOE – European Polysaccharide Network of Excellence – is a research, education and knowledge transfer network connecting Companies, Academic and Research Institutions working on, or interested in, polysaccharides. It was established with help from the European Commission. EPNOE's main missions are to offer a networking platform enabling close interactions between members in order to favour innovation, boost knowledge transfer and organize R&D activities in a totally confidential manner; to organize basic and applied research for the study of fundamental concepts, the testing of new ideas and the development of new products based on or containing polysaccharides; to organize education in polysaccharide science.

IUPAC The International Union of Pure and Applied Chemistry is the world authority on chemical nomenclature and terminology, including the naming of new elements in the periodic table; on standardized methods for measurement; and on atomic weights, and many other critically-evaluated data. A neutral and objective scientific organization, IUPAC was established in 1919 by academic and industrial chemists who shared a common goal – to unite a fragmented, global chemistry community for the advancement of the chemical sciences via collaboration and the free exchange of scientific information. Throughout its long history IUPAC has fulfilled that goal through the creation of a common language and the standardization of processes and procedures.

J&K Scientific is committed to being much more than a 'chemical company'. Since our inception in 1992, we have been guided by a clear mission: To Accelerate Scientific and Industrial Development, Thereby Serving Humanity. With our exceptional staff team and state-of-the-art production facilities, we offer our customers high quality chemicals and related products and services to aid in drug discovery, environmental protection and myriad other endeavors that serve humanity

LC Scientific Inc. is an R&D laboratory and small-scale production facility located just north of Toronto in Canada, specializing in efficient synthesis of established and novel compounds for the chemical, biochemical, bio-energy, and therapeutic research fields. We are a team of Carbohydrate Synthesis Specialists quickly gaining worldwide recognition for innovative process development and resourceful problem solving on the toughest carbohydrate and glycoconjugate projects.

Megazyme is a global leader in the development, manufacture and supply of analytical reagents, enzymes and assay kits for the food, feed, dairy and wine industries. Megazyme is unique in being the only company worldwide specifically developing and supplying diagnostic test kits and reagents for the beverage, cereals, dairy, food, feed, fermentation, biofuel and wine industries. From the company's inception as a manufacturer of test kits and reagents, Megazyme has become a globally recognised leader in the development of new testing methods for carbohydrates and enzymes for the cereals and related industries. Megazyme International was founded in 1988 by Barry McCleary, CEO

Mizutani Foundation for Glycoscience is a nonprofit organization established on October 1, 1992. The aim of the Foundation is to promote worldwide progress of glycoscience, especially that of glycoconjugates. The mission of Mizutani Foundation is to contribute to the welfare of mankind through promotion of basic studies on glycoconjugates mainly by distributing research grants to qualified scientists regardless of their nationalities or ethnic origins. The Foundation may also support international conferences on glycoscience to facilitate exchanges between Japanese and foreign glycoscientists and carry out other activities that are vital for the progress of glycoconjugate-related sciences.

Omicron Biochemicals, Inc. develops new and improved methods to introduce stable isotopes, primarily 13C, 2H, 15N and 18O, into biologically-important compounds. We specialize in the synthesis of labeled carbohydrates, nucleosides, and their derivatives, but will undertake the synthesis of other classes of compounds on a custom basis. Our particular strengths lie in the preparation of site-specifically labeled compounds, but we are also capable of producing uniformly labeled products. The approaches used in synthesis include chemical, chemi-enzymic, enzymic and biological methods. We also collaborate with other laboratories as needed to develop new products or distribute our products. We take pride in providing quality stable isotopes in support of a wide range of scientific research programs worldwide. Our research activities frequently result in publication in peer-reviewed scientific journals, either independently or in conjunction with other research groups.

Organic & Biomolecular Chemistry (OBC) is a weekly journal for the publication of highly significant original research and reviews in all areas of organic chemistry, including organic synthesis, physical organic chemistry, supramolecular chemistry and bioorganic chemistry. OBC's scope includes, but is not limited to the following: Total synthesis, synthetic methodology, synthesis of biomolecules, bioorganic chemistry, as well as organic and bioorganic aspects of the following: Supramolecular and macromolecular chemistry, theoretical and computational chemistry, catalysis, medicinal chemistry, natural products, including biosynthetic aspects.

ProZyme, Inc. was founded in 1990 with a vision of establishing a high-quality, value-added and customer-driven biochemical reagent company. We are proud, 25 years on, to continue to provide the highest level of products, services and support to our customers at large and small biotechnology, biopharmaceutical and academic institutions across the globe. ProZyme's flagship product line is our glycobiology portfolio, alongside streptavidin, phycobiliprotein and conjugate offerings. Our range of glycoanalysis products was enhanced in 2003 with the purchase of Glyko, Inc., including the biochemical reagent line of Oxford GlycoSciences (OGS). ProZyme has maintained a commitment to invest in, develop, release and support products in the exciting and rapidly-expanding area of glycobiology. ProZyme's scientists are also able to perform custom analytical services, and we welcome the opportunity to collaborate with our customers.

Springer Throughout the world, we provide scientific and professional communities with superior specialist information – produced by authors and colleagues across cultures in a nurtured collegial atmosphere of which we are justifiably proud. We foster communication among our customers – researchers, students and professionals – enabling them to work more efficiently, thereby advancing knowledge and learning. Our dynamic growth allows us to invest continually all over the world. Springer is the publisher of Cellulose, Analytical & Bioanalytical Chemistry, and Journal of Materials Science.

Southern Regional Research Center (SRRC) in New Orleans, Louisiana, is part of the United States Department of Agriculture's (USDA) Agricultural Research Service (ARS). The mission of these centers, strategically located around the United States, is to lead America toward a better future through agricultural research and information. ARS research is organized into National Programs that benefit the American consumer by improving the quality of agricultural output and finding new uses for agricultural products. Current areas of research at SRRC include the chemical modification of cotton for value added applications; cotton-based nonwovens; developing technologies that enable growth and profitability in the commercial conversion of sugarcane, sweet sorghum and enery beets into sugar, advanced biofuels and bioproducts; improving the product value of catfish; increasing the value of cottonseed; and others.

Sussex Research is a well-known specialist in carbohydrate synthesis with an established international reputation. Founded in 1996, Sussex Research is a carbohydrate-enabling company, excelling in the design, synthesis and characterization of carbohydrate-based molecules. Our unique product portfolio enables researchers across a variety of scientific disciplines to readily explore and exploit glycobiology via facile introduction onto small molecules, biological molecules, nanoparticles or surfaces. Our advanced capabilities in carbohydrate synthetic chemistry, purification and characterization has benefited our collaborators in academia, research institutions, biotechnology and pharma at all stages of research & development from basic research to clinical trials. Sussex Research is highly active in custom synthesis, contract manufacture and contract research & development of oligosaccharides, glycosaminoglycans, glycoamino acids, glycopeptides, modified carbohydrates with functionalized linkers/spacers and a wide variety of synthetic building blocks.

TCI – Tokyo Chemical Industry Co., Ltd is a leading worldwide manufacturer of specialty organic chemicals founded in 1946. TCI provides organic laboratory chemicals as well as pharmaceutical, cosmetic and functional materials. More than 60 years of synthesis experience and multi-purpose plants enable TCI to offer more than 25,000 products as well as custom synthesis. TCI established overseas facilities in North America, Europe, China and India to serve customers worldwide.

V-LABS, INC., was established in 1979 as an independent, private organization. V-LABS, INC, provides carbohydrates, polysaccharides, and glycosidic enzymes used in research in the food, biochemical, and pharmaceutical industries, as well as in universities and research institutions. We have developed molecular weight standards of water soluble polysaccharides. V-LABS, INC., also offers analytical service for carbohydrates and polysaccharides, including gas chromatography, high performance liquid chromatography, ultraviolet and infrared spectroscopy. We also determine residual solvents and off-flavors in oilseed products, feeds, packaging materials, and plastics. We have studied extensively the development of rancidity in oils and oil-containing products such as nuts and grains.

Wyatt Technology is the recognized leader in light scattering instrumentation and software for determining absolute molar mass, size, charge and interactions of macromolecules and nanoparticles in solution. Wyatt provides in-line multi-angle static light scattering - SEC-MALS; field flow fractionation (separation with no stationary phase)- FFF-MALS; composition gradients for interaction analysis - CG-MALS; high-throughput dynamic light scattering-DLS; high-sensitivity electrophoretic mobility-MP-PALS; differential refractometry, and differential viscosity. Wyatt's training, service and support are widely considered the industry benchmark of excellence.

Yamazen Corporation is Japan's prominent manufacturer of the Single & Dual Channel Automated Flash Systems (Preparative Liquid Chromatographs) and high resolution columns, with 35 years of technical experience in flash chromatography. Expert in chromatography! Yamazen Science Inc is a sales and service operation for the US business.

General Information

About New Orleans
The city of New Orleans has a storied past beginning with its establishment on the banks of the Mississippi River by Jean-Baptiste Le Moyne de Bienville in1718. In 1763, New Orleans passed from French rule to Spanish rule. In 1800 the city reverted back to French rule, and in 1803 was sold to the United States by Napoleon as part of the Louisiana Purchase. The original city is now known as the French Quarter or Vieux Carré, still a thriving commercial and residential neighborhood. The French Quarter is both a lively tourist destination and a genuine residential neighborhood. Although referred to as the French Quarter, the architecture is actually Spanish, characterized by ornate wrought iron balconies festooned with flowers and cool, interior patios. New Orleans is known around the world for its food, music and *joie de vivre*.

Weather in New Orleans in July
The weather in July in New Orleans is warm and humid. The average high daytime temperature is 32°C / 90°F, but it can go higher. The average low temperature (night time and early morning) is 24°C / 75°F. Afternoon rain and clouds can help to cool things off. Daytime humidity ranges from 56% to 82%.

What to Wear
All buildings in New Orleans are heavily air conditioned in the summer, so you will want to have a light jacket, wrap or sweater for indoors, just in case you get too chilly. Light, casual clothing is appropriate for outdoors.

Daytime Tours
Most of our Wednesday afternoon tours will spend some time outdoors. We recommend light clothing for all outdoor activities, as well as a hat. We also recommend liberal use of sun block. Hydration is important during hot weather, and we will make sure that plenty of water is available for our guests. Check the description of each tour in case there are other instructions.

Grocery Store Near By: Rouses (a Louisiana supermarket chain) is at 701 Royal St., open until 1:00 AM, with a little bit of everything.

Drugstores Near By: CVS at 800 Canal Street and Walgreens, 900 Canal Street are open 24 hours.

Shopping Centers in the Area:
The Shops at Canal Place, 333 Canal Street, is home to many exclusive stores, featuring upscale clothing and home items, located in the center of a four-star hotel, The Westin New Orleans Canal Place.

Shops at Jax Brewery, 600 Decatur Street, in the French Quarter, are located in an 1890s brewery that has been converted into a shopping center with apparel, accessory & gift stores.

The Outlet Collection at Riverwalk, 500 Port of New Orleans, is an outlet mall located on the Mississippi River featuring many national retailers and an extensive food court overlooking the river.

Other Interesting Places Nearby and Things to Do:
Preservation Hall, authentic New Orleans jazz in a primitive setting, 726 St. Peter St. Reserved seats $35 to $45, http://preservationhall.tunestub.com/eventList.cfm. There's an old trombone case hanging over the street for a sign.

Take a carriage ride around the French Quarter. Mule-driven, open carriages are available for hire on the Decatur side of Jackson Square.

Have coffee and beignets at the French Market; afterwards, browse the small shops, booths and flea markets in the area. The Café du Monde is across Decatur St. from Jackson Square

Stroll Jackson Square: This is the heart of the French Quarter, with the St. Louis Cathedral, the Cabildo and the Presbytère on its eastern edge, the French Market and Mississippi River on the west; and the Pontalba Apartments, the first apartments in the United States on the north and south sides of the square. The Presbytère is a unit of the Louisiana State Museum, showcasing an elaborate and exquisite collection of Mardi Gras artifacts and memorabilia. Hours are Tuesday - Sunday, 10 a.m. to 4:30 p.m.; closed Monday. $6.00 for adults; $5.00 for seniors

Visit **The Historic New Orleans Collection**, 533 Royal Street, is a museum, research center and publisher dedicated to the study and preservation of the history of New Orleans and the Gulf South region of the United States. Open Tuesday through Saturday 9:30 AM to 4:30 PM; Sunday 10:00 AM to 4:30 PM; closed Monday. http://www.hnoc.org/

Visit the **New Orleans Pharmacy Museum** at 514 Chartres Street, between St. Louis and Toulouse Streetsi n the heart of the French Quarter. Admission is $5.00 Open Tuesday to Saturday 10am-4pm; closed on Sunday and Monday. This is the home of America's first licensed pharmacist. http://www.pharmacymuseum.org/

Meeting Venue
All technical events will be held at the Marriott Hotel, 555 Canal Street. The hotel is located on the edge of the French Quarter, making it very convenient to numerous interesting sites.

Airport Shuttle: Airport Shuttle is the official provider of transportation to the Louis Armstrong International Airport. A one-way trip is $24 per person, and reservations should be made 24 hours in advance. The phone number is +1(504) 522-3500. The link for the service (the preferred contact) is XVIII International Carbohydrate Symposium Official Discounted Airport Shuttle

Registration: An onsite registration and information desk will be maintained throughout meeting hours.

Speakers and Session Chairs: Speakers please check with your session chair in your meeting room at least 10-15 minutes before the start of the session. Session chairs are requested to adhere strictly to the schedule in order for participants to plan attendance at different parallel sessions. If a speaker fails to show, please hold the time open until the next scheduled speaker.

Information for Presenters: Presenters may check their presentations in the speaker-ready room, Galerie 1. Please bring your presentation to the speaker-ready room at least 24 hours before the start of your session so it can be put onto the Symposium network. The speaker-ready room will be open Sunday through Thursday from 8:00 AM to 5:00 PM.

Lunch: Lunch will be available in the Carondelet Room (third floor) on Monday, Tuesday and Thursday for delegates who have purchased lunch tickets. The meals will feature typical everyday Louisiana fare. Lunch tickets may be purchased onsite. Please purchase a day or more ahead.

Information for Poster Presenters: Posters will be located in Preservation Hall and will be up throughout the meeting. To encourage and facilitate poster viewing, posters will be co-located with the Exhibits and where coffee will be served. Posters may be mounted in Preservation Hall on Monday, July, 18, 10.00 AM onwards and removed on Thursday July 21st, 3:40 PM onwards. Poster presenters will officially attend their posters either on Monday evening (Poster Session 1, 5:50 pm-7:00 pm) or Tuesday evening (Poster Session 2, 5:20pm-6:40 pm), as shown in the program.

Exhibitors

Commercial exhibits representing a range of carbohydrate-related businesses and products will be co-located in Preservation Hall along with the posters and coffee breaks to maximize opportunities for networking.

Currency Exchange for Visitors and Tipping

Many of the ICS attendees are experienced visitors to the US and I do not presume to advise them, since I travel the other direction. For those who are not experienced international travelers, the following may be of help. Note that when you arrive, the Customs and Border Protection Officer may ask you to prove that you have sufficient funds to cover your costs during your visit to the US. So, be prepared to show your hotel receipt and a credit card.

The USA uses "plastic" (credit, debit, and ATM cards) for almost everything. At the same time, there is not much capacity for changing money. You just do not see equivalents of signs for Cambio or Wechsel when walking down the street. My best advice is to bring and use a credit card that has a small or no "foreign transaction fee." Also, bring at least a small amount ($50 to $100) of US Dollars from home. For more cash, have an ATM or debit card tied to your checking account to meet any local expenses that require cash. Another option is a "Prepaid Dollar Card" available in some countries. All that said, the Marriott Hotel will change many currencies for guests, up to $100 per day. The lobby has a Chase Bank ATM. The nearby Canal Place Shopping Center (333 Canal St.) has a Capital One Bank ATM. The Hancock/Whitney Bank, 228 St. Charles Avenue, can handle larger transactions during business hours. Further advice is available at https://www.tripadvisor.com/Travel-g191-s601/United-States:Banks.And.Money.html.

Lonely Planet gives the following advice about tipping in the US:

"Tipping is not optional; only withhold tips in cases of outrageously bad service."
Bartenders and Restaurant servers – 15% to 20%; Hotel Maids - $2 to $4 per night
Taxi drivers - 10% to 15%

Recording presentations is not permitted.

It is strictly prohibited to use any device (e.g., camera phone, camera, tape or digital recorder) to capture images or sounds, or stream, upload or rebroadcast speakers or presentations without written consent from the 28th ICS Organizers.

Marriott Hotel

Second Floor Plan

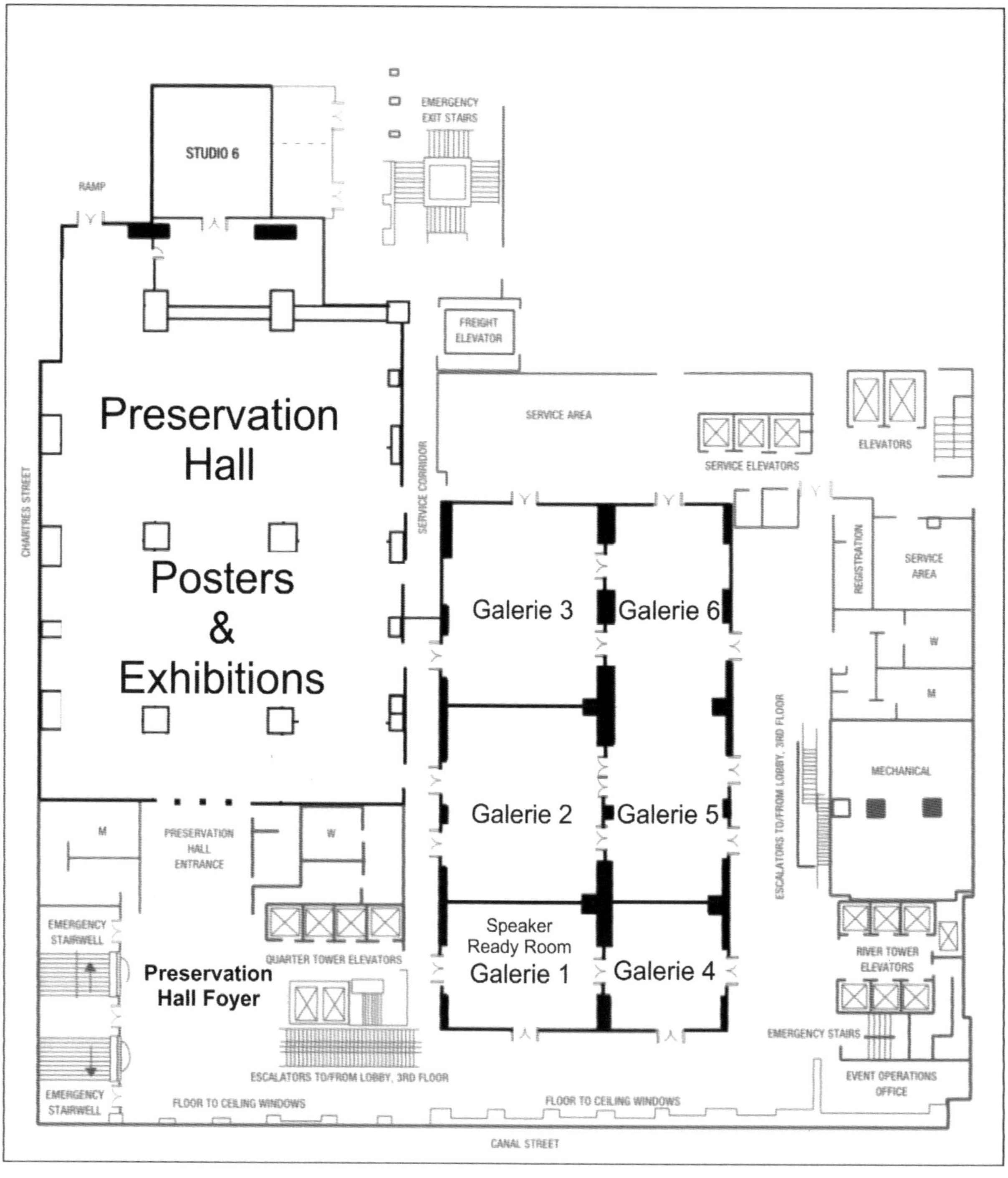

Marriott Hotel

Third Floor Plan

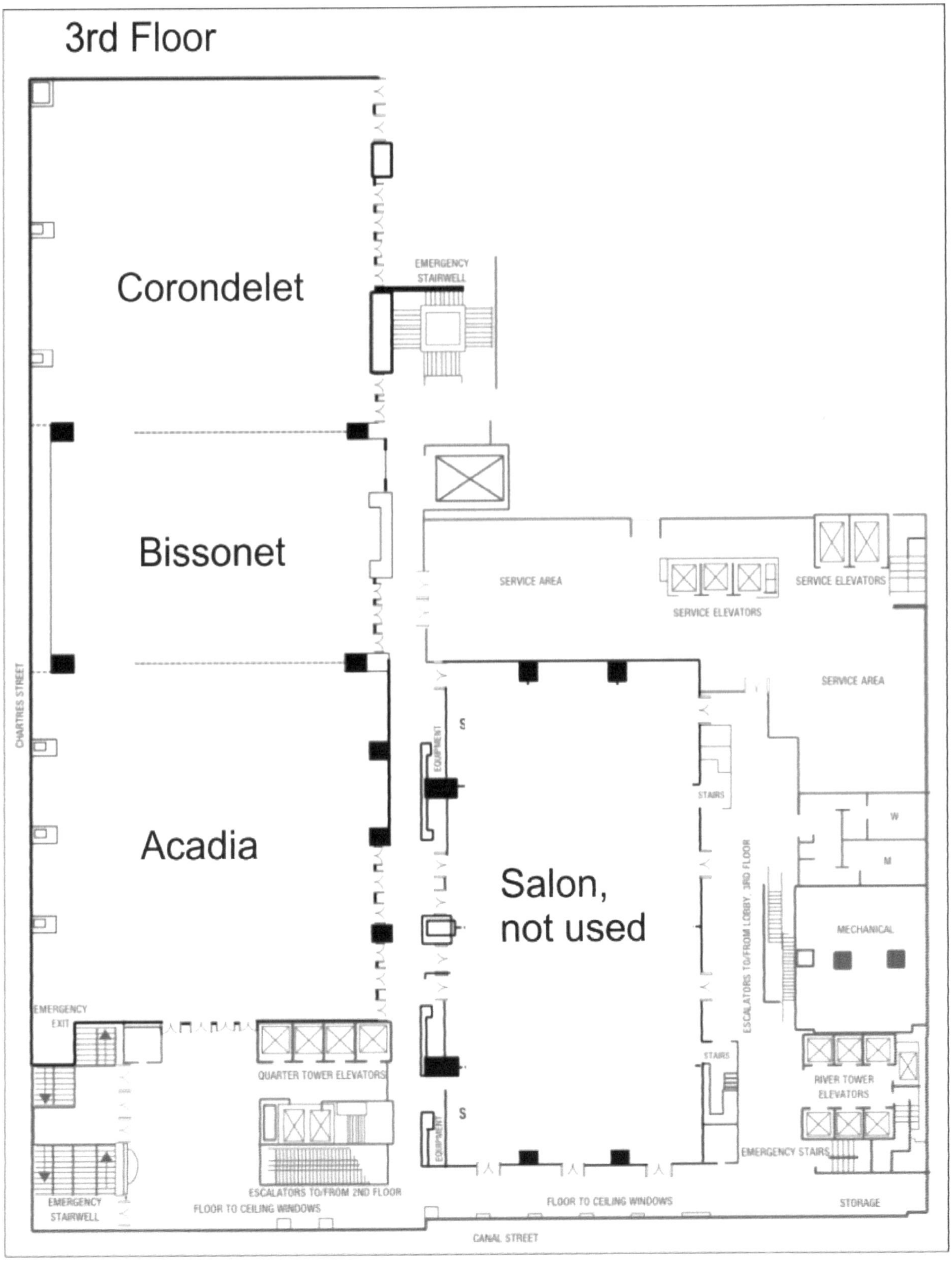

EVENTS

PRE-MEETING PANEL DISCUSSION: DIRECTIONS IN GLYCOSCIENCE

Under the leadership of Dr. Serge Perez, former Scientific Director of the European Synchrotron Radiation Facility and former Head of the CERMAV Research Institute and author of "A Roadmap for Glycoscience in Europe", Dr. Pamela Marino, Chief, Division of Pharmacology, Physiology and Biological Chemistry, National Institutes of Health, and Dr. Koichi Kato, Project Leader, National Institutes of Natural Sciences, Japan, there will be a special panel discussion, 10:00 am on Sunday, July 17, concerning the various carbohydrate/glycoscience roadmaps developed in recent years.

OPENING CEREMONY

The meeting will officially begin with the Opening Ceremony at 4:00 PM, with welcoming words from Alfred French, the Symposium Chair, Dr. Pamela Marino, United States National Institutes of Health, and Dr. N. Jayaraman, President of the International Carbohydrate Organization. Professor Johannes Vliegnethart will represent the International Union of Pure and Applied Chemistry, and Dr. French will close with an overview of the symposium. Following the Opening Ceremony, the first session of the meeting will begin with the Whistler Award address from Prof. Benjamin Davis. Dr. Benjamin Swarts, recipient of the ICO Young Researcher Award, will give his award address, followed by the first Plenary Lecture to be given by Prof. Carolyn Bertozzi.

WELCOME RECEPTION

Registrants are invited to attend the welcome reception that follows the close of the First Session, Sunday evening at 7:30 PM.

CULTURAL NIGHT at ST. LOUIS CATHEDRAL

In keeping with ICS tradition, Tuesday evening, at 7:00 PM, attendees are invited to a Cultural Night at the historic St. Louis Cathedral, situated in the heart of the French Quarter. Here, you will be treated to a selection of music typical of New Orleans. The Cathedral is a short walk from the Marriott Hotel.

WEDNESDAY AFTERNOON EXCURSIONS

Wednesday afternoon and evening are set aside for excursions, including an afternoon Jazz Steamboat Cruise, Houmas House and Garden tour and a Creole Plantation tour. If you prefer, you can see the Haunted side of New Orleans or take a ride on an airboat in a Louisiana swamp. There is limited available seating for two special tours – the Turtle Cove Wetlands Research Center and the Domino Sugar Refinery. A City Tour will be available Sunday noon, ahead of the meeting, and there will be a French Quarter Walking Tour and Cooking Class on Monday for guest attendees.

CLOSING CEREMONY

The Closing Ceremony will begin at 5:20 PM Thursday. Included will be awarding of the best poster awards, an invitation from Dr. Antoni Planas to the 19th European Carbohydrate Symposium in Barcelona in 2017 and an invitation from Professor Amelia Rauter to the 29th International Carbohydrate Symposium in Lisbon in 2018. Dr. Alfred French will conclude with final remarks.

CLOSING RECEPTION AND BANQUET at MARDI GRAS WORLD

After the closing ceremony on Thursday, starting at 6:40 PM, buses will take delegates to Mardi Gras World for a tour of the facility where the the elaborate floats used in the Mardi Gras celebrations are designed, produced, and stored, followed by a reception and banquet. The reception will be outside by the Mighty Mississippi River. The Symposium Banquet will follow the reception. This promises to be a wonderful evening, full of fantasy and fun.

Roy L. Whistler Award

The International Carbohydrate Organization is delighted to announce that the Roy L. Whistler International Award in Carbohydrate Chemistry for 2016 has been awarded to Professor Benjamin G. Davis, Professor of Chemistry at the University of Oxford, UK.

In 1984, the International Carbohydrate Organization established the Award in honor of Professor Roy L. Whistler, to recognize scientists who have made contributions of excellence in carbohydrate chemistry and biochemistry and with promise of continuing significant contributions. The Award is recognized with a plaque, a check for US $25,000, and an invitation to present the opening lecture at the 28th International Carbohydrate Symposium (ICS).

Ben Davis studied chemistry at the University of Oxford where he was awarded a B.A. (1993) and a D.Phil. (1996) with George Fleet. After a postdoctoral fellowship with Bryan Jones at the University of Toronto and a lectureship at the University of Durham, he joined the Dyson Perrins laboratory at the University of Oxford with a Fellowship at Pembroke College, and was promoted to full Professor in 2005.

Prof Davis' research defines a broad arc of applied and fundamental carbohydrate research in the areas of Synthetic and Chemical Biology and Chemical Medicine. His group has developed a toolbox of reactions and techniques for post-expression modification of proteins that extends and expands the biomolecules available through molecular biology techniques. These have been applied to synthesize a wide range of novel glycoproteins and glycoconjugates that have been used to address important questions in biology. His group has pioneered a range of synthetic carbohydrate methods including carbohydrate asymmetric catalysis, carbohydrate-inhibitor arrays, complex oligosaccharide synthesis, and has delivered new understanding into the anomeric effect, enzyme mechanism, and complex glycoconjugate biosynthesis. A crowning achievement of his research is the discovery of methods enabling C-C bond formation with proteins, most notably protein olefin cross-coupling and Pd-mediated protein cross-coupling.

In recent years Prof Davis has extended his studies to include increasing sophisticated synthetic constructs such as a synthetic glucosylated nucleosome, and complex virus- and cell-like synthetic assemblies. Outcomes of his work in molecular medicines include new methods for the diagnosis of TB and novel vaccine strategies targeting HIV and bacterial pathogens. His group has also pioneered a range of carbohydrate-based and targeted non-invasive imaging methods for disease diagnosis.

Prof Davis has published over 200 papers, a book (Carbohydrate Chemistry, with Antony Fairbanks), more than 30 patents, and has founded two companies (Glycoform, Oxford Contrast). He is a recipient of the RSC Corday-Morgan Medal, and was elected Fellow of the Royal Society in 2015.

ICO Young Researcher Award

The International Carbohydrate Organization is delighted to announce that the Young Researcher Award for 2016 has been awarded to Dr Benjamin Swarts, Central Michigan University, USA.

The International Carbohydrate Organization established the Award in 2014 to recognize outstanding international achievements on any aspect of glycochemistry and/or glycobiology. The Award is recognized with a plaque, an invitation to present at the upcoming International Carbohydrate Symposium (ICS), and covers registration and accommodation expenses.

Ben Swarts (1982) studied chemistry at the College of Wooster, Ohio, and completed a PhD (2010) at Wayne State University with Professor Zhongwu Guo. He undertook postdoctoral studies (2010-13) with Prof Carolyn Bertozzi at UC Berkeley, supported by American Cancer Society and Center for Emerging and Neglected Diseases Frameworks for Global Health postdoctoral fellowships. In 2013, he joined the Department of Chemistry and Biochemistry at Central Michigan University, where he is an Assistant Professor.

Dr. Swarts has made significant contributions to the development of new chemoenzymatic methods for the synthesis of complex carbohydrates and, more recently, to chemical glycobiology. In his PhD studies, he developed new methods for the chemical synthesis and enzymatic modification of protein glycolipid anchors, the glycosylphosphatidylinositols. During his postdoctoral work he established an innovative method for the study of the mycobacterial trehalome, the cohort of glycoconjugates containing trehalose, and helped to develop new super-sensitive fluorophores for the discrimination of mycobacteria through detection of sulfatase activity. His present work seeks to expand the chemical biology toolkit for studying glycolipids and other cell envelope components of the mycobacteria, which include the pathogen that causes tuberculosis. Dr Swarts' recent work in this area is highlighted by the development of a one-step chemoenzymatic synthesis of trehalose-based probes for the detection of mycobacteria, and the creation of new bioorthogonal chemical reporters that facilitate the study of glycolipids and proteins in the mycobacterial outer membrane.

Dr Swarts has published 17 papers, including 7 from his work at CMU. His research is supported by a Research Corporation Cottrell College Scholar Award, and the National Institutes of Health.

Claude S. Hudson Award

The Claude S. Hudson award was established by the American Chemical Society Carbohydrate Chemistry Division in 1946 to recognize outstanding contributions to carbohydrate chemistry in education, research, or applications.

To celebrate the 2015 Hudson Award with the international glycoscience community, the Carbohydrate Chemistry Division of the American Chemical Society has decided to move the award symposium from National ACS meeting to the ICS2016.

The 2015 winner of the Hudson Award is Geert-Jan Boons, Distinguished Professor in Biochemical Sciences at the Complex Carbohydrate Research Center (CCRC) of the University of Georgia.

Dr. Boons received his M.Sc. in Chemistry in 1987 and his Ph.D. in Synthetic Carbohydrate Chemistry in 1991 from the State University of Leiden in The Netherlands. Prior to joining the faculty at the CCRC in 1998, he spent seven years in the United Kingdom, first as a postdoctoral fellow at Imperial College, London, and the University of Cambridge, and then as a lecturer and professor at the University of Birmingham. In 2003, Dr. Boons was awarded the Carbohydrate Research Award for Creativity in Carbohydrate Science by the European Carbohydrate Association. Also in 2003, he was elected chairman for the 2005 Gordon Research Conference on Carbohydrates. In 2004, Dr. Boons received the Horace Isbell Award by the Division of Carbohydrate Chemistry of the American Chemical Society and was appointed Franklin Professor of Chemistry in the College of Arts and Sciences at the University of Georgia. In 2012, he received the Creative Research Inventor's Award by the University of Georgia Research Foundation and in 2013 was appointed UGA Foundation Distinguished Professor in Biochemical Sciences. He was awarded the Roy L. Whistler International Award in Carbohydrate Chemistry in 2014 by the International Carbohydrate Organization.

Our Plenary Speakers

Carolyn R. Bertozzi

"Glycocalyx engineering toward probing cancer glycome evolution"

Carolyn Bertozzi is the Anne T. and Robert M. Bass Professor of Chemistry and Professor of Chemical & Systems Biology and Radiology (by courtesy) at Stanford University, and an Investigator of the Howard Hughes Medical Institute. Her research interests span the disciplines of chemistry and biology with an emphasis on studies of cell surface glycosylation pertinent to disease states. She completed her undergraduate degree in Chemistry from Harvard University in 1988 and her Ph.D. in Chemistry from UC Berkeley in 1993. After completing postdoctoral work at UCSF in the field of cellular immunology, she joined the UC Berkeley faculty in 1996. In June 2015, she joined the faculty at Stanford University coincident with the launch of Stanford's ChEM-H institute. She is the Editor-in-Chief of *ACS Central Science*, which recognizes chemistry's central role in bridging the physical and life sciences and is the recipient of many honors and awards, including Elected Fellow of the National Academy of Inventors, Elected Member of the National Academy of Sciences and the Whistler Award in Carbohydrate Chemistry.

Vincent Bulone

"A journey into the world of Eukaryotic cell walls: structure and biosynthesis of essential polysaccharides"

Professor Vincent Bulone holds a PhD in Biochemistry from the University of Lyon, France (1992). After two post-doctoral positions in Melbourne and Oslo, he successively became Lecturer at the University of Grenoble, Professor of Biochemistry at the University of Lyon and Professor of Plant Glycobiology at the Royal Institute of Technology (KTH) in Stockholm, Sweden. Vincent Bulone is currently the Head of the Department of Glycoscience at KTH and the Director of the Australian Research Council (ARC) Centre of Excellence in Plant Cell Walls headquartered at the University of Adelaide, Australia. His research activities comprise fundamental investigations on cell wall polysaccharide structure and metabolism in higher plants, oomycetes and fungi, with particular interest in the characterization of membrane-bound complexes with glycosyltransferase activity; engineering of plant and microbial polysaccharide biosynthetic pathways; quantitative proteomics; and engineering and characterization of carbohydrate-based composites. The knowledge generated from his fundamental research is translated into applications relevant to the following areas: Bioenergy, Biomaterials, Biorefinery, Food Sustainability and Disease Control. In 2010 he received the Bruce Stone Award for Excellence in Plant Cell Wall Biochemistry in recognition of his seminal work on cellulose biosynthesis and cell wall biochemistry in higher plants and eukaryotic microorganisms.

Bruce Hamaker

"Carbohydrate quality and how the concept may relate to healthier foods"

Dr. Hamaker is the Director of the Whistler Center for Carbohydrate Research, and Distinguished Professor, Purdue University, West Lafayette, Indiana. He received his Ph.D. in Food Chemistry from Purdue University in 1986, followed by post-doctoral research at the Instituto de Investigación Nutriciona/Johns Hopkins University, in Lima, Peru. From 1988-1992 he was Assistant Professor in the Department of Food Science, University of Arkansas, after which he returned to Purdue University and became affiliated with the Whistler Center. He has been the director of the center since 2003. Professor Hamaker's general research areas include carbohydrates and health, slowly digestible starch, dietary fiber and cereal chemistry. In 2008, Professor Hamaker received the International Award from the Institute of Food Technologists and in 2012 the WK Kellogg International Food Security Award from the Institute of Food Technologists. He has been active in the Institute of Food Technologists, American Association of Cereal Chemists, American Chemical Society and the American Society of Nutrition and is the author of over 200 publications.

Koichi Kato

"Structural views of glycan-dependent determination of glycoprotein fates in cells"

Koichi Kato received his Ph.D. in 1991 at Graduate School of Pharmaceutical Sciences, the University of Tokyo under the supervision of Prof. Yoji Arata, and continued his research as Assistant Professor and Lecturer in the same institution. In 2000, he moved to Nagoya City University as full Professor. In 2008, he moved to his current position at Okazaki Institute for Integrative Bioscience, National Institutes of Natural Sciences, holding a Professor position concurrently at Nagoya City University. His research interests include structural analyses of glycoconjugates by NMR spectroscopy and other biophysical and biochemical methods. Based on his accomplishments, he was awarded the Pharmaceutical Society of Japan Award for Young Scientists in 2000, the Pharmaceutical Society of Japan Award for Divisional Scientific Promotions as well as the 48th Baelz Prize in 2011. He currently acts as project leader of JSPS Grant-in-Aid for Scientific Research on Innovative Areas "Dynamical ordering of biomolecular systems for creation of integrated functions".

Muthia Manoharan

"Carbohydrate conjugated RNAi therapeutics"

Dr. Muthiah Manoharan serves as Senior Vice President of Innovation Chemistry at Alnylam Pharmaceuticals, Cambridge, Massachusetts, USA. Dr. Manoharan joined Alnylam in 2003. He built the chemistry group at Alnylam and pioneered the discovery of GalNAc conjugated siRNAs for RNA interference (RNAi) based human therapeutics. He was the former Executive Director of Medicinal Chemistry at Isis Pharmaceuticals, Inc., a leading biotechnology company focused on antisense oligonucleotide-based therapeutics where he had a12-year tenure. With a distinguished career as a world-leading nucleic acid and bioconjugate chemist, Dr. Manoharan is an author of nearly 200 publications and over 300 abstracts, as well as the inventor of over 190 issued U.S. patents. Prior to Isis Pharmaceuticals, He earned his Ph.D. in chemistry at the University of North Carolina-Chapel Hill and conducted post-doctoral work at Yale University and the University of Maryland. He was the recipient of the M. L. Wolfrom award of the ACS Carbohydrate Chemistry Division in 2007.

James C. Paulson

"Sialic acids as determinants of self"

James C. Paulson obtained his PhD (Biochemistry) in 1974 from the University of Illinois at Champaign-Urbana, and did post-doctoral work at Duke University Medical Centre, in Durham, North Carolina from 1974-78. From 1978 – 1990 he rose from Assistant Professor to full Professor and vice-chair in the Department of Biological Chemistry at the UCLA School of Medicine where he developed an interest in analysis of receptor specificity of influenza viruses from different host species. From 1990-1999 he served as Vice President and Member Board of Directors of Cytel Corporation, La Jolla, CA. From 1999-present he has been Professor, in the Departments of Cell and Molecular Biology, Chemical Physiology, and Immunology and Microbial Sciences at The Scripps Research Institute, in La Jolla, California. He served as acting President & CEO from 2014-2015 and is currently Cecil and Ida Green Professor and Chair of Cell and Molecular Biology. His current research interests include the roles of glycan binding proteins in the modulation of immune cell signaling, and the receptor specificity of mammalian and animal influenza viruses. For a listing of publications go to: http://www.scripps.edu/mb/paulson/publications.html

Soledad Penadés

"Glycans and glyconanotechnology"

Professor Soledad Penadés is Principal Researcher Glyconanoparticles, Centro de Investigación Cooperativa en Biomateriales, at CIC biomaGUNE, Guipúzkoa, Spain. She obtained her PhD in chemistry from University Complutense, Madrid, and did postdoctoral research at the Max-Planck Institut für Strahlenchemie and the University of Frankfurt. From 1976-1979 she was Associate Professor at Medizinische Hochschule, Hannover. From 1980-1982 she was visiting scientist at the Max-Planck Institut fúr Kohlenforschung, Mülheim. In 1979 she became Associate Researcher of the Spanish Research Council (CSIC) at the Institute for Organic Chemistry in Madrid. In 2003 she was appointed Research Professor at the Institute for Chemical Research in Seville. During these years she developed a new area of research focused on supramolecular chemistry and carbohydrate recognition. She developed model systems (glycophanes) to study the intermolecular forces involved in carbohydrate recognition in water, with particular emphasis on carbohydrate self-recognition. She participated actively in the creation of the Institute for Chemical Research in the new Research Centre of La Isla de la Cartuja (CICIC), in Seville, moving there in 1997 when she was Vice-Director until 1999. In recent years her laboratory has been developing new methodologies to obtain highly polyvalent carbohydrate surfaces (glyconanoparticles) based on metallic nanoclusters to study carbohydrate-to-carbohydrate interactions.

Stephen G. Withers

"Accessing new and improved CAZymes through metagenomics and directed evolution"

Stephen Withers is Khorana Professor of Chemistry and Biochemistry, University of British Columbia, Vancouver, Canada and served as the founding Director of the UBC Centre for High-Throughput Biology (CHiBi). He received his Ph.D. from Bristol University in 1977 under M.L. Sinnott and did postdoctoral work at Alberta University from 1977-1982 under N.B. Madsen and B.D. Sykes. Among his many awards and recognitions are: Corday Morgan Medal of the Royal Society of Chemistry (1990); Rutherford Medal, Royal Society of Canada (1992), Charmian Medal of the Royal Society of Chemistry, U.K., (2001), Whistler Award, ICS (2002), Fellow of the Royal Society of Canada (2002), R.U. Lemieux Award for Research in Organic Chemistry (2011); Fellow of the Royal Society of London (2012), Centenary Prize Royal Society of Chemistry (2012), CIC Medal (2016). His research interests focus primarily on mechanisms of enzymes that catalyze glycoside formation and hydrolysis, since these play crucial roles in all areas of biology. Applications of his research range from the development of new catalysts for industrial processes to the design, synthesis and testing of new therapeutics. He has over 480 publications.

PROGRAM AT A GLANCE		
Date	**Time**	**Event**
Sunday, July 17		**Pre-Meeting Event**
	10:00 AM - Noon	Research Directions in Glycoscience
		ICS-2016
	9:00 AM - 7:00 PM	Registration & Information
	12:00 Noon	ICO Committee Meeting & Lunch
	4:00 PM	Opening Ceremony
	4:30 PM	Roy L. Whistler Award Lecture
	5:15 PM	Coffee Break
	5:45 PM	ICO Young Researcher Award Lecture
	6:15 PM	Plenary Lecture 1 - Carolyn R. Bertozzi
	7:30 PM	Welcome Reception
Monday, July 18	7:30 AM - 7:00 PM	Registration & Information
	8:30 AM	Plenary Lecture 2 - James C. Paulson
	9:15 AM	Plenary Lecture 3 - Muthiah Manoharan
	10:00 AM	Coffee Break
	10:30 AM	4 Keynote Lectures / 5 Concurrent Sessions
	11:00 AM	Technical Program / 5 Concurrent Sessions
	12:20 PM - 2:00 PM	Lunch / Exhibits & Poster Viewing
	2:00 PM	4 Keynote Lectures / 5 Concurrent Sessions
	2:30 PM	Technical Program / 5 Concurrent Sessions
	3:50 PM	Coffee Break / Exhibits & Poster Viewing
	4:20 PM	4 Keynote Lectures / 5 Concurrent Sessions
	4:50 PM	Technical Program / 5 Concurrent Sessions
	5:50 PM - 7:00 PM	Exhibits & Posters Session 1
Tuesday, July 19	7:30 AM - 6:00 PM	Registration & Information
	8:30 AM	Plenary Lecture 4 - Soledad Penadés
	9:15 AM	Plenary Lecture 5 - Koichi Kato
	10:00 AM	Coffee Break / Exhibits & Poster Viewing
	10:30 AM	4 Keynote Lectures / 5 Concurrent Sessions
	11:00 AM	Technical Program / 5 Concurrent Sessions
	12:20 PM - 1:30 PM	Lunch / Exhibits & Poster Viewing
	1:30 PM	4 Keynote Lectures / 5 Concurrent Sessions
	2:00 PM	Technical Program / 5 Concurrent Sessions
	3:20 PM	Coffee Break / Exhibits & Poster Viewing
	3:50 PM	4 Keynote Lectures / 5 Concurrent Sessions
	3:20 PM	Technical Program / 5 Concurrent Sessions
	5:20 PM - 6:40 PM	Exhibits & Poster Session 2
	7:00 PM	Cultural Event at St. Louis Cathedral

Wednesday, July 20	7:30-3:00	Registration & Information
	8:30 AM	Plenary Lecture 6 - Stephen G. Withers
	9:15 AM	Hudson Award Lecture
	10:00 AM	Coffee Break / Exhibits & Poster Viewing
	10:30 AM	4 Keynote Lectures / 5 Concurrent Sessions
	11:00 AM	Technical Program / 5 Concurrent Sessions
	12:40 PM	Excursions
Thursday, July 21	7:30 AM - 7:00 PM	Registration & Information
	8:30 AM	Plenary Lecture 7 - Vincent Bulone
	9:15 AM	Plenary Lecture 8 - Bruce Hamaker
	10:00 AM	Coffee Break / Exhibits & Poster Viewing
	10:30 AM	4 Keynote Lectures / 5 Concurrent Sessions
	11:00 AM	Technical Program / 5 Concurrent Sessions
	12:20 PM - 2:20 PM	Lunch / Exhibits & Poster Viewing
	2:20 PM	Technical Program / 5 Concurrent Sessions
	3:40 PM	Coffee Break
	4:00 PM	Technical Program / 5 Concurrent Sessions
	5:20 PM	Closing Ceremony
	6:40 PM	Buses leave for Banquet
	7:00 PM	Receptiona & Banquet at Mardi Gras World

Abstracts of Our Award Winners

Roy L. Whistler Award

Benjamin G. Davis
University of Oxford Chemical Research Lab, Oxford, United Kingdom

Sugars and proteins

This lecture will cover emerging areas in our group in the use of chemical probes & modulators of carbohydrate processing and signalling events. Delineation of mechanisms of carbohydrate-processing systems and their ligands allows not only an understanding of their fundamental role in Biology and Immunology but also the use of molecules to modulate and manipulate such processes, with a strong associated potential for diagnosis, therapy, application and intervention in Medicine and other biotechnologies.

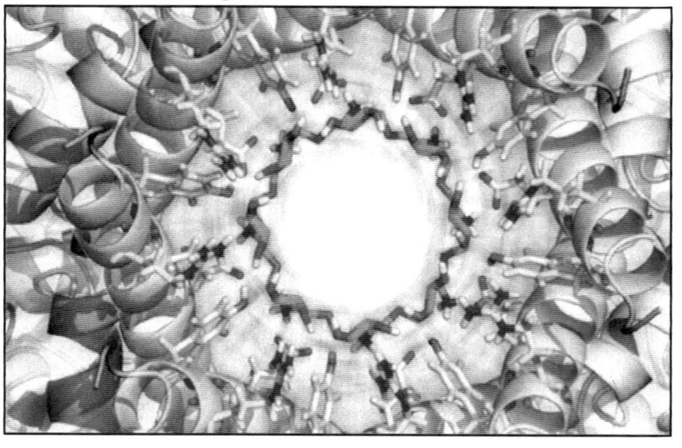

Claude S. Hudson Award

Geert-Jan Boons
Complex Carbohydrate Research Center, University of Georgia, Athens, GA, United States

An integrated approach to uncover ligands for heparan sulfate binding proteins

Although hundreds of heparan sulfate binding proteins have been identified, and implicated in a myriad of physiological and pathological processes, very little information is known about ligand requirements for binding and mediating biological activities by these proteins. This difficulty results from a lack of technology for establishing structure activity relationships, which in turn is due to the structural complexity of natural HS and difficulties of preparing well-defined HS-oligosaccharides. To address this deficiency, we have developed an integrated approach for the identification of ligands for HS-binding proteins employing affinity purification, sequencing of HS-oligosaccharides by novel mass spectrometric approaches and the chemical synthesis of libraries of HS-oligosaccharides for structure activity relationships. The methodology has been employed to identify ligands for the ROBO-1 which is a protein involved in cell signaling and development. Furthermore, the synthetic HS-oligosaccharides have been used for the developed of a microarray and screening efforts have uncovered novel aspect of heparan sulfate-protein binding.

ICO Young Researcher Award

Benjamin Swarts
Central Michigan University, Mount Pleasant, Michigan, United States

Marking the mycomembrane: Chemical tools for studying and targeting glycolipids of the mycobacterial outer membrane

Mycobacteria, including the global pathogen *Mycobacterium tuberculosis*, possess a unique outer membrane—the mycomembrane—which is composed of an array of distinctive glycolipids, including the trehalose- and arabinogalactan-linked mycolates. Despite the importance of mycomembrane components to mycobacterial physiology and pathogenesis, tools for investigating their structures, dynamics, and contributions to disease progression are limited. Chemical probes, such as radiotracers and bioorthogonal chemical reporters, are valuable tools for tagging and investigating cell surface components, and often can be used in situ. Our lab's objective is to develop mycobacteria-specific probes that facilitate the study of the mycomembrane and that may provide new opportunities to target it for tuberculosis diagnosis and treatment. This talk focuses on the development and applications of probes based on trehalose, a disaccharide that is central to mycomembrane biosynthesis but absent from humans.

Abstracts of Our Plenary Speakers

Carolyn R. Bertozzi
Department of Chemistry, Stanford University, Stanford, California, United States

Glycocalyx engineering toward probing cancer glycome evolution

While altered glycosylation patterns have long been identified as hallmarks of cancer, their functional significance with respect to tumor progression are not well understood. Two examples are overexpression of mucin glycoproteins (densely glycosylated cell-surface molecules with unusual physical properties) and hypermodification of glycoproteins with the terminal sugar sialic acid. These glycosylation phenotypes are found on numerous cancer types with highly varied underlying driver mutations and their magnitude tends to correlate with tumor aggressiveness. To test hypotheses regarding the functional significance of cancer glycomes, we developed an approach to engineer the cell surface "glycocalyx" with chemically defined glycopolymers that emulate cancer-associated structures. Using living polymerization and chemoselective ligation chemistries, we synthesize glycopolymers functionalized with a biophysical probe on one end and a lipid capable of membrane insertion on the other. These biomimetic structures can be displayed on live cell membranes where they acquire functions analogous to natural mucin glycoproteins. Our work using this platform suggests that hypersialylation protects tumor cells from innate immune surveillance, whereas mucin upregulation alters the physical properties of the glycocalyx so as to promote focal adhesion formation and signaling. Both glycophenotypes can thus be understood as evolutionary adaptations of cancer cells under various selective pressures.

Vincent Bulone
Department of Glycoscience, Royal Institute of Technology, Stockholm, Sweden; ARC Centre of Excellence in Plant Cell Walls, University of Adelaide, Adelaide, Australia

A journey into the world of eukaryotic cell walls: Structure and biosynthesis of essential polysaccharides

Eukaryotic cell walls are complex extracellular matrices that largely consist of polysaccharides. Despite strong differences in molecular composition and architecture across taxonomic groups, cell walls exert similar functions regardless of the type of organism considered. The biosynthesis of cell wall polysaccharides is one of the most important biochemical processes in plant biology and microbiology. Indeed, cell wall polysaccharides are essential for normal development of plant cells, fungi and other eukaryotic microorganisms, and play major roles in cell integrity and defense responses to various stresses. Of high biological importance are the biosynthetic processes that take place at the plasma membrane, such as cellulose, chitin and (1→3)-β-D-glucan synthesis. The latter polysaccharides are linear homopolymers that each contain a single type of glycosidic linkage. But this apparent simplicity hides sophisticated molecular biosynthetic events that have been unexpectedly challenging to decipher. This presentation will review and compare the structures and compositions of major classes of cell walls from higher plants, oomycetes and fungal species. Our latest results on cellulose, chitin and (1→3)-β-D-glucan biosynthesis in model organisms will be presented, and the corresponding biochemical steps will be linked to polysaccharide structure and crystallization into microfibrils. Our recent and current research efforts will be illustrated by the use of multidisciplinary approaches, including determination of cell wall composition and polysaccharide structure, biochemistry of membrane-bound proteins, *in vitro* synthesis of carbohydrate polymers and their characterization with chemical and biophysical methods, and functional characterization of recombinant carbohydrate synthases from plants and pathogenic oomycetes. The limitations of these approaches will be discussed together with possible strategies to better understand the biosynthesis of microfibrillar cell wall polysaccharides in eukaryotic cells and their assembly.

Bruce Hamaker
Purdue University, West Lafayette, IN, United States

Carbohydrate quality and how the concept may relate to healthier foods

What does "carbohydrate quality" mean, and could it have a similar differentiating effect on food selection for good health as say omega-3 fatty acids for fats and oils? Food carbohydrates can be generally separated into available and nonavailable carbohydrates; the first consisting of digestible carbohydrates or free sugars that can be absorbed into the body (glucose, fructose, galactose) and the latter indigestible carbohydrates comprising dietary fiber. The term "quality" currently refers mainly to digestion rate of glycemic carbohydrates with the idea that glycemic response profiles should be moderated, but location of digestion and absorption is also important because it can activate enteroendocrine cells found in the small intestine ileum whose hormones control physiological processes affecting food motility (mainly gastric emptying rate) and appetitive response and even food intake. There are a number of ways to affect carbohydrate digestion rate and location starting with the chemical structure of starch and linkage patterns of alpha-linked oligo-glucans. Amylose in cooked gelatinized foods is well known to retrograde quickly on cooling and to have resistant properties, but also amylopectin fine structures with higher proportion and length of intermediate and long linear chains tend to retrograde faster and form aggregates with somewhat slow digestion rate. Di- and oligosaccharides with unusual alpha-linkages have been shown to have moderated digestion and glycemic profiles. In an effort to delay digestion to the ileum, we have also studied the use of non-competitive food-based phenolic inhibitors of the mucosal alpha-glucosidases to moderately delay digestion to the ileum for triggering appropriate gut hormones. The concept of improving carbohydrate quality of foods, and strategies to achieve this end will be presented.

Koichi Kato
1. Okazaki Institute for Integrative Bioscience and Institute for Molecular Science, National Institutes of Natural Sciences, Okazaki, Japan; 2. Graduate School of Pharmaceutical Sciences, Nagoya City University, Nagoya, Japan

Structural views of glycan-dependent determination of glycoprotein fates in cells

In the endoplasmic reticulum (ER), the nascent proteins are modified with a triantennary high-mannose-type-oligosaccharide, which embeds glycotopes as protein-fate determinants. The oligosaccharides displayed on the proteins are enzymatically trimmed by a series of glycosidases in the ER, resulting in exposure of the determinants, which are recognized by the intracellular lectins operating as ER chaperones, cargo receptors, and degradation mediators.

During these fate-determination processes, glucosidase II cleaves the non-reducing terminal glucose residue of the N-glycan intermediates and thereby creates the monoglucosyl glycoforms, which are capable of interacting with the ER chaperones. Interestingly, this enzyme also attacks its produced monoglucosylated glycoproteins, sequestering them from the chaperones. The nonglucosyl high-mannose-type-glycans bearing the terminal trimannosyl structure serve as signals for recognition by certain types of cargo receptors. Glycoproteins yet to obtain their native structures are tagged by terminal glucosylation on the oligosaccharide so as to bring them into additional chaperone-assisted folding process. The selective tagging of the glycoproteins with structural defect is committed by an ER folding sensor enzyme UGGT.

We have characterized the molecular recognition mechanisms of the N-glycan processing enzymes and the intracellular lectins by structural biology approaches employing X-ray crystallography and other biophysical techniques. We have also developed a paramagnetism-assisted NMR technique in conjunction with computational approach for delineating dynamic conformational ensembles of the high-mannose-type oligosaccharides carrying the protein-fate determinants. Our structural data have provided mechanistic insights into the glycan-dependent determination of glycoprotein fates in cells.

Muthiah Manoharan
Alnylam Pharmaceuticals Inc, Cambridge, MA, United States

GalNAc-conjugated oligonucleotides as a new paradigm in RNAi therapeutics for human diseases

Synthetic small interfering RNAs (siRNAs) act as therapeutic agents through the RNA interference (RNAi) pathway and are specific and potent inhibitors of gene expression. These agents may be designed to target disease pathways previously considered "undruggable". Recently, delivery of therapeutic siRNAs to liver hepatocytes upon subcutaneous administration has been achieved by conjugating chemically modified siRNAs with multivalent N-acetylgalactosamine (GalNAc) residues that are recognized by the asialoglycoprotein receptor (ASGPR). siRNA-GalNAc conjugates efficiently target and silence disease-causing genes produced in liver hepatocytes in animal models and humans. Using this conjugation platform,
Alnylam is advancing several RNAi agents specific for liver targets through preclinical and clinical development to address diseases with unmet medical need. Our progress with the chemistry of siRNA-GalNAc conjugates and applications in several therapeutic areas will be presented.

James C. Paulson, Shiteng Duan, Matthew S. Macauley, Ryan McBride, Corwin Nycholat, Lijuan Pang, Amrita Srivastava, Robert de Vries
Departments of Cell and Molecular Biology, Chemical Physiology and Immunology and Microbial Sciences, The Scripps Research, La Jolla, CA, United States

Sialic acids as determinants of self

Sialic acids are displayed on the surface of all cells as terminal sugars on glycans of glycoproteins and glycolipids. As such, they are prominent determinants of self, recognized as receptors by numerous microbes as host cell receptors, and by glycan binding proteins of the immune system that regulate immune cell reactivity. Siglec receptors (**si**alic acid-binding immuno**g**lobulin-type **lec**tins), comprising 14 members in humans and 9 members in mice are predominately expressed on immune cells. Many of the Siglecs help the immune system distinguish between self and non-self, and carry regulatory motifs that can regulate receptors involved in mediating immune responses. With a goal of elucidating the natural roles of Siglecs and exploiting them to regulate immune cell responses, we are developing glycan ligands of high affinity and selectivity for each Siglec. Nanoparticles displaying such ligands can effectively target Siglec expressing cells in vivo, providing a modular platform for delivering cargo or modulating immune responses of the targeted cell. We have demonstrated the utility of Siglec ligand-decorated nanoparticles to deliver chemotherapeutic agents to lymphoma cells, deliver antigens to antigen presenting cells, and to suppress antigen-mediated immune responses of B cells and mast cells, demonstrating the potential for targeting Siglecs to modify immune cell mediated disease (NIH grants AI050143, AI099141, HL107151, CA013889).

Soledad Penadés
CIC biomaGUNE, San Sebastián, Spain

Glycans and glyconanotechnology

To study and intervene in carbohydrate-mediated interactions our laboratory has developed an integrated chemical-nanotechnology strategy (Glyconanotechnology) to produce sugar functionalized gold nanoclusters with multivalent carbohydrate display (glyconanoparticles, GNPs). Glyconanotechnology has also allowed the preparation of biofunctional nanoparticles with fluorescent and/or magnetic properties for application to cellular labelling and imaging. The preparation of multifunctional GNPs incorporating a selected set of ligands (DNA, RNA, peptides, fluorescence probes, etc.) on the same cluster is comprised within the potential of this technology. In this lecture, I will review this glyconanotechnology strategy and will discuss some applications to basic studies in carbohydrate-mediated interactions as well as to biomedical problems.

Stephen G. Withers
Chemistry, University of British Columbia, Vancouver, BC, Canada

Accessing new and improved CAZymes through metagenomics, synthesis and directed evolution

Gaining access to enzymes that are able to degrade or synthesise specific glycosides, especially those containing rare or unnatural sugars, can be very challenging. In theory this could be achieved by redesign of the active site of existing enzymes, but very few such approaches have been successful in producing enzymes that are of practical value. Likewise, modification of selectivity via directed evolution approaches has not been as fruitful as had been hoped. Both such approaches could benefit from access to more and better "start points" for modification.

Mother Nature has already supplied an enormous diversity of biocatalysts, each of which could serve as a starting point for directed evolution studies. The problem can be in accessing this diversity in a reasonably efficient manner. Here I shall describe two ways in which we have approached this

problem. One of these is through synthesis of libraries of genes from within promising glycoside hydrolase families, selected to represent phylogenetically distinct subfamilies of genes. This will be illustrated with our work using a library of over 100 GH1 glycosidases. In the other approach we have used activity-based, or functional metagenomics to generate a library of over 600 expressed glycosidases. These libraries have been subjected
to "first-pass" high-throughput characterisation for substrate specificity, thermal stability, pH profile and mechanism. This sub-library has then been screened to identify preferred catalysts for cleavage of specific unnaturally modified sugars (e.g. azido sugars). Once candidate enzymes are identified they are then mutated to "glycosynthase" versions that can be used to "tag" glycans. In parallel we have screened the large library for enzymes that can be used to remove the Gal or GalNAc residues that function as the antigenic determinants from A and B type red blood cells, thereby generating "universal" O type blood. Such libraries can also be used to select the optimal candidate for further improvements through directed evolution.

Abstracts of Our Keynote Speakers

Kawthar Bouchemal
University Paris Sud, Chatenay Malabry, France.

Supramolecular polysaccharide assemblies: A novel therapeutic strategy against mucosal microbial infections

Generally, systemic administration of drugs is the first intention for the prevention and/or the treatment of mucosal microbial infections. However, this administration route leads to drug toxicity, drug resistance and low drug efficacy or even inactivity, due to the low dose available on mucosal tissues. In the other hand, the efficacy of local drug delivery is limited by dilution with physiological fluids and low residence time close to mucosal tissues.

In this context, tailor-made polysaccharide conjugates able to self-organize in aqueous solution in the form of supramolecular assemblies were specifically designed for the prevention and/or the treatment of severe mucosal microbial infections, particularly: candidiasis, trichomoniasis, leishmaniasis, and viral infections with HIV-1 and HSV-2.

When local administration on mucosal surfaces is foreseen, the assemblies should be formulated into a vehicle specifically designed for mucosal administration. The idea exposed here is to design multifunctional formulations composed of polysaccharide assemblies dispersed into mucoadhesive and thermosensitive hydrogel. Our data showed that the resulting formulations exhibited high antimicrobial activities even without adding any drug. When associated with antimicrobial drugs, the assemblies and/or the hydrogel act as a "booster" allowing to the active concentration of the drug to be reduced. Our translational and interdisciplinary approach will lead to design more active and better tolerated formulations against mucosal microbial infections.

Bouchemal K, Aka-Any-Grah A, Dereuddre-Bosquet N, Martin L, Lievin-Le-Moal V, Le Grand R, Nicolas V, Gibellini D, Lembo D, Poüs C, Koffi A, Ponchel G. *Antimicrob. Agents. Chemother*. 2015, *59*, 2215-2222.
Lembo D, Donalisio M, Laine C, Cagno V, Civra A, Bianchini E.P, Zeghbib N, Bouchemal K. *Eur. J. Pharm. Biopharm*. 2014, *88*, 275-282.
Pradines B, Bories C, Vauthier C, Ponchel G, Loiseau P. M, Bouchemal K. *Pharm. Res*. 2015, *32*, 1229-1236.
Galus A, Mallet JM, Lembo D, Cagno V, Djabourov M, Lortat-Jacob H, Bouchemal K. *Carbohydrate Polymers*. 2016, *136*, 113-120.
Pradines B, Djabourov M, Vauthier C, Loiseau P. M, Ponchel G, Bouchemal K. *Colloids and Surfaces B: Biointerfaces*. 2015, *135*, 669-676

Adam B. Braunschweig, Yeting Zheng, Yujia Zhou
Department of Chemistry, University of Miami, Coral Gables, FL, United States

Synthetic C lectins: A new route for targeting cell-surface glycans

Cell-surface glycans — including glycoproteins, glycolipids, and glycopolymers — are involved in cell differentiation, cell proliferation, the immune response, and many other biological processes, and thus are tantalizing targets for drug delivery, disease detection, and therapies. Despite the promise of small molecules that specifically recognize cell-surface glycans — termed synthetic lectins — for various biomedical applications, it still remains very challenging to develop receptors with the requisite affinity and specificity for targeting or therapy. Typically, synthetic lectins are designed following Cram's principles of complementarity and preorganization, whereas natural lectins — carbohydrate binding proteins — are typically promiscuous and flexible, and high affinity and selectivity is achieved by accessing cooperativity, multivalency, and sometimes using calcium ions to bridge H-bonds (C-lectins). We report a bioinspired synthetic lectin that achieves high-affinity recognition in water by accessing biomimetic binding modes (Figure A), whose selectivity is altered in the presence of calcium ions. Binding between this new class of synthetic lectins and monosaccharides was analyzed by ^{1}H NMR (Figure B), and these studies suggests several new approaches for the design of carbohydrate receptors.

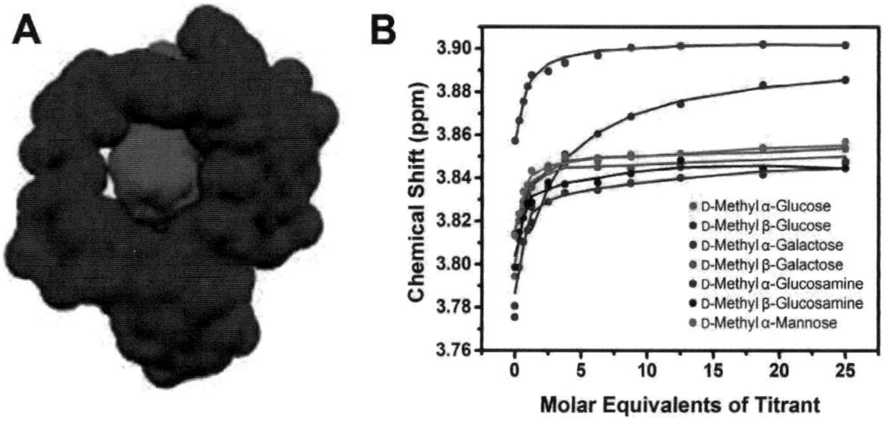

(A) Model of synthetic C-Lectin binding MeO-β-D-Galactose. (B) Curves showing the changes in chemical shifts of receptor ^{1}H NMR peaks in H_2O upon titrating in various monosaccharide guests.

Xing Chen
College of Chemistry and Molecular Engineering, Peking University, Beijing, China

Liposome-assisted metabolic glycan labeling *in vivo*

Mammalian glycan structures and patterns vary in different tissues and in distinct cell types. We are interested in probing glycosylation *in vivo* with cell selectivity and tissue specificity. Toward this goal, we have developed a liposome-assisted bioorthogonal reporter (LABOR) strategy, which exploits liposome nanoparticles encapsulating azidosugars. By outfitting liposomes with targeting ligands, we selectively delivered azidosugars to specific cells of interest, which allowed for cell-selective metabolic labeling and visualization of sialylated glycans in co-culture. Furthermore, in a xenograft mouse model, we applied this strategy for targeted imaging and proteomic analysis of tumor-associated glycans. Finally, the LABOR strategy enabled, for the first time, *in vivo* metabolic labeling and fluorescent imaging of brain sialylation in living mice.

Gene Cioffi
Pharmacology, University of South Alabama, Mobile, AL, United States
Functional glycomics: Our sugarlined blood vessels

Endothelial cells line the inside of blood vessels forming a semipermeable barrier across which macromolecules and solutes are selectively transported. Indeed, in the lung, inflammation-induced endothelial barrier disruption promotes accumulation of plasma proteins in the interstitial and alveolar compartments leading to impaired gas exchange. An essential component of both pulmonary and systemic endothelial barrier structure and function is the glycocalyx. It is known that the glycocalyx of endothelial cells isolated from the pulmonary artery (PAECs) exhibits differences compared to the glycocalyx of pulmonary capillary endothelial cells (PMVECs), which supports the phenomenon of endothelial heterogeneity.

We questioned how the glycocalyx of endothelial cells isolated from the systemic circulation is similar or different from that of the pulmonary circulation and the brain circulation. Further, we wanted to extend studies into the glycocalyx differences within a specific circulatory system (*i.e*, artery *vs.* capillary). We evaluated extracellular N-and O-linked carbohydrate expression and intracellular O-GlcNAc expression in PAECs, PMVECs, aortic (AOECs), and brain microvascular (BMVECs) endothelial cells using Click Chemistry.

Specific carbohydrates and linkages were probed using lectin binding assays and mass spectrometry. Total and free sialic acids were measured; contributions from lycosaminoglycans (GAGs) were evaluated. Finally, functional contributions of the glycocalyx to endothelial barrier integrity were measured using Electrical Cell Substrate Impedance Sensing (ECIS). As an example of our preliminary findings, we observed that AOECs express greater amounts of N-and O-linked carbohydrates on their cell surfaces; cells from the larger vessels (AOECs and PAECs) express more intracellular O-GlcNAc compared to other cells. Lectin and mass spectrometry analyses revealed significant differences in carbohydrate expression, branching patterns and linkages between all cell types. Collectively these results reveal that the glycocalyxes of pulmonary, systemic, and brain endothelial cells express unique glycan signatures.

Catherine E. Costello
Center for Biomedical Mass Spectrometry, Boston University School of Medicine,
Boston, MA, United States
The dynamic state of glycan mass spectrometry

Co-or post-translational modifications (PTMs) affect protein-protein and cell-cell interactions, localization and turnover rates, as well as changing the physicochemical properties of the nascent biopolymer. Yet these are often overlooked as important factors that regulate biological activity; glycans and other PTMs are frequently removed prior to structural analysis, either because their heterogeneity complicates molecular assignments or because they are used as "handles" to enrich a representative subset within the biological sample and are then discarded to release the captured analytes. Methods that retain the modifications, on the other hand, enable examination of their site-specific distributions and detailed structures, in order to reach conclusions about structure-activity relationships. High resolution separations, pre-and post-ionization, *e.g.*, multistage chromatography and ion mobility, respectively, can deliver purified analytes to the mass analyzer and high performance mass analysis coupled with varied dissociation methods, especially the new electron-activated options, can provide quite unequivocal assignments. In order to become widely used, the methods need optimization, the assembly of relevant databases and automation of assignments; these processes are ongoing. The current status will be discussed with illustrations from studies that explore the human glycoproteome in health and disease and the microbiome of pathogenic and nonpathogenic microbes. The BUSM CBMS is supported by NIH grant P41 GM104603, and several S10 equipment grants. Collaborative projects are supported by NIH and Foundation grants.

Mary K. Cowman
Tandon School of Engineering, New York University, New York, NY, United States

Hyaluronan and hyaluronan fragments

The glycosaminoglycan hyaluronan (HA) is a key component of the microenvironment surrounding cells. In healthy tissues, HA molecules have extremely high molecular mass, and consequently large hydrodynamic volumes. Tethered to the cell surface by clustered receptor proteins, HA molecules crowd each other, as well as other macromolecular species. This leads to severe non-ideality in physical properties of the biomatrix, because steric exclusion leads to an increase in effective concentration of the macromolecules. In our model for mutual crowding of polymers, the excluded volume depends on both polymer concentration and hydrodynamic volume / molecular mass. The biomechanical properties of the extracellular matrix, tissue hydration, receptor clustering, and receptor-ligand interactions are strongly affected by the presence of HA, and its molecular mass. In inflammation, reactive oxygen and nitrogen species fragment the HA chains. Depending on the rate of chain degradation relative to the rates of new synthesis and removal of damaged chains, short fragments of the HA molecules can be present at significant levels. Not only are the physical properties of the extracellular matrix affected, but the HA fragments de-cluster their primary receptors and act as endogenous danger signals. We have developed multiple bioanalytical methods to isolate and quantify HA fragments for potential use in medical diagnostic tests. We are also collaborating on projects to develop and apply biotherapeutic agents to modulate signaling by HA fragments, for new therapeutic approaches to wound healing and cartilage repair without fibrosis.

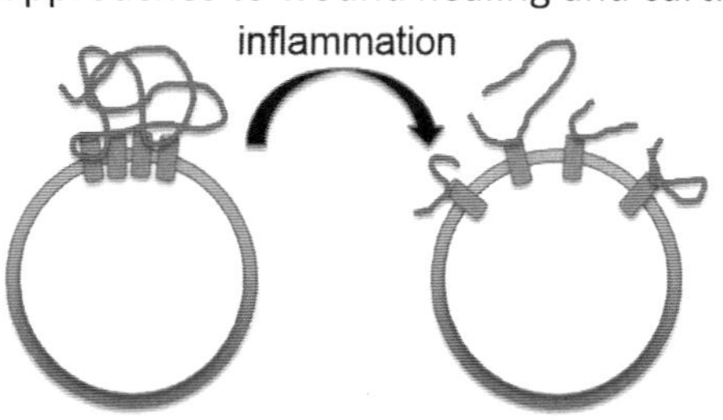

Cristina De Meo
Chemistry, Southern Illinois University Edwardsville, Edwardsville, IL, United States

The effect of O-substituents on the outcome of chemical sialylation: H-Bond-mediated aglycone delivery and beyond

Sialic acid-containing glycoconjugates are natural products that, in the last two decades, have become important synthetic targets, due to their direct involvement in numerous biological phenomena. Unfortunately, the stereoselective synthesis of the glycosidic linkage between sialic acid and a glycoconjugate chain is still a synthetic challenge. Among many sialyl donors that have been investigated, those modified at C-5 have been the most exploited.[1] On the other hand, whereas the effect of O-protecting groups on the outcome of common glycosylation reactions is well known, no systematic study of O-modifications on sialylation has yet emerged.

As a part of a program to investigate the effect of O-protecting groups on chemical sialylation, herein we report our study of sialyl donors modified at O-4 and O-7 positions. Our recent findings on the existence of a conformational equilibrium suggest that these positions may be relevant in sialylation reactions.[2] That is because the all-axial conformation of the oxacarbenium intermediate places O-7 and O-4 in a close proximity to the anomeric center. Having investigated the effect of electronically and sterically demanding substituents[3], our more recent study has refocused on the investigation of 4-O- and/or 7-O-picoloylated sialyl donors. In accordance with the H-bond-mediated aglycone delivery (HAD), a concept recently reported by Demchenko at al., [4] these sialyl donors are expected to deliver the incoming nucleophile from the α-face of the ring. The effect of the C-5 amine protecting group in cooperation with the HAD concept has also been investigated.[5]

(1) De Meo, C.; Priyadarshani, U. Carbohydr. Res. 2008, 343, 1540-1552.
(2) Meo, C. D.; Wallace, C. E.; Geringer, S. A. Org. Lett. 2014, 16, 2676-2679.
(3) Premathilake, H. D.; Gobble, C. P.; Pornsuriyasak, P.; Hardimon, T.; Demchenko, A. V.; De Meo, C. Org. Lett. 2012, 14, 1126-1129.
(4) Yasomanee, J. P.; Demchenko, A. V. J. Am. Chem. Soc. 2012, 134, 20097-20102.
(5) De Meo, C; Escopy, S.; Geringer, S. Najafi F. 2016, in preparation.

Stephen J. Eichhorn
Physics, University of Exeter, Exeter, United Kingdom

Cellulose in water and other solvents-aligning and self-assembling

Cellulose has a remarkable affinity to water, which makes it one of the "greenest" polymers in existence compared to many other manmade materials. However, cellulose does not dissolve in water, and many other common solvents, which is a both curiosity and issue for its processing outside of natural systems. The recalcitrance of cellulose to solvents will be explored in this talk, but also the potential for its self-assembly in a colloidal state and through other solvents into fibrous structures. Remarkably, with very little chemical or mechanical "persuasion" highly oriented fibres can be formed. Finally, some recent work on the use of ionic liquids to process high performance cellulose fibres will be presented.

Beat Ernst
Department of Pharmaceutical Sciences, University of Basel, Magden, Switzerland

The conformational variability of FimH – which conformation represents the therapeutic target?

Urinary tract infections (UTI) belong to the most common infections worldwide and are predominantly caused by uropathogenic *E. coli* (UPEC). UPEC initiate the infection cycle by adhering to urothelial cells through the bacterial lectin FimH. Located at the tip of type 1 pili, FimH mediates adhesion by binding to high-mannose glycoproteins present on urothelial cells. Preventing the initial adhesion by blocking FimH with glycomimetic antagonists represents an alternative to the conventional antibiotic treatment of UTI, which has become increasingly inefficacious due to the aggravating issue of antibiotic resistance.

The FimH-mannose interaction is peculiar in that it is shear stress dependent, meaning that FimH mediates weaker binding at low shear stress and stronger binding at high shear stress. This type of bond is called catch-bond and results from the ability of FimH to adopt two distinct bound conformations, a medium- and a high-affinity conformation. In the urinary tract, the shear stress-induced switch from the low- to the high-affinity conformation becomes an effective tool for UPEC to evade clearance by the bulk flow of urine.

So far, efforts in developing FimH antagonists have been predominantly focused on the high-affinity conformation, and the importance of the conformational heterogeneity of FimH has been mostly neglected. To this effect, we have investigated FimH binding of various clinically and functionally relevant FimH variants in a cell-based assay under hydrodynamic conditions. Clinical FimH variants were typically characterized by a catch-bond behavior and hence, as being conformationally dynamic. By

studying bacterial binding in a cell motility assay, we were able to show that conformationally dynamic FimH variants are highly mobile on a
mannosylated surface, allowing optimal colonization in the bladder. In addition, by switching to the high-affinity conformation, these bacteria can resist upcoming shear stress and are thus protected from being washed out through urination. In contrast, a variant with constitutive high-affinity formed long-lived interactions with mannose in the absence of shear stress, rendering bacteria immobile. It is evident that dynamic FimH variants are pathophysiologically favored and represent the dominant therapeutic target. To further elucidate this conformational issue, antagonists with high binding affinities to both affinity states were synthesized and characterized in various *in vitro* and *in vivo* models.

Jovyn K. Frost[1], Bernadine Flanagan[2], Michael Gidley[2], Douglas Rosendale[1], David Brummell[1], John Monro[1]
1. The New Zealand Institute for Plant & Food Research, Palmerston North, New Zealand
2. ARC Centre of Excellence in Plant Cell Walls, Centre for Nutrition & Food Sciences, University of Queensland, Brisbane, QLD, Australia

Potato digestibility: Composition, structure and function of cell walls

Low glycemic impact potatoes are a healthy source of carbohydrates as they offer a slower rate of glucose release, providing sustained energy levels. In this study, we screened a New Zealand-based potato germplasm collection for glycemic response using *in vitro* methods. We identified three advanced breeding lines that showed an effect of significantly reduced glycemic impact in the presence of their cell walls, compared with hydrolysis of their isolated starch alone. Starch properties known to affect digestibility such as amylose content and degree of amylopectin branching did not differ significantly; therefore we set out to investigate the influence of the cell walls on digestibility in these potato lines.

Monosaccharide composition analysis confirmed the galactan-rich nature of potato cell walls. Glycan microarray profiling revealed a difference in the solubility profiles of pectic side chains galactans and arabinans, with a higher relative abundance of these epitopes extracted in strong alkali in the potato lines with lower glycemic effect. Solid state ^{13}C NMR spectroscopy revealed interactions between pectic galactan and cellulose in the cell wall preparations of these lines. Potato lines containing elevated amounts of these pectic galactans also showed an increased resistance to pectin solubilization during cooking. Findings suggest that possession of galactan-enriched rhamnogalacturonan-1 cell walls may provide a sturdier wall that better resists solubilization during cooking, and effectively impedes enzyme access for starch hydrolysis in an *in vitro* model. We further investigated the outcome of the undigested polysaccharides using an *in vitro* fermentation model. This presentation will discuss the role of plant cell walls and a shift in its composition in the context of dietary consequences.

Matthew I. Gibson
University of Warwick, Coventry, United Kingdom

Glycosylated nanoparticles for the detection of pathogens and toxins

Carbohydrate-protein interactions are inherently weak but also, in the case of monosaccharides, very nonspecific with most lectins being capable of binding a vast array of carbohydrates. In Nature the precise 3-D presentation of branched glycans gives rise to specificity, which allows viral or bacterial docking to host cells. The reproduction of these specific multivalent arrays offers huge promise as the development of new anti-adhesives and pathogen sensors. However, the synthesis of native glycans is often challenging, and more accessible methods to reproduce their recognition functions but with reduced synthetic burden are required. We have developed glycosylated gold nanoparticles which give a colourmetric response upon lectin binding (red to blue). Crucial to this is engineering the gold particle interface using synthetic polymers to ensure a balance between rapid aggregation upon lectin binding (to give colour change) but sufficient colloidal stability to prevent false positives. RAFT polymerization is used as it generates a thiol (for gold binding) at one end, and the installation of a bio-orthogonal group at the other end for glycan capture. We have developed multiplexed sensors to 'barcode' lectins (such as cholera toxin or ricin) using multiple nanoparticle 'channels' providing discrimination but using only easy to access sugars. This talk will cover key synthetic concepts and highlight recent success with this methodology.

References:

Richards, SJ., Jones, MW, Hunaban, M., Haddleton, DM., Gibson, M.I. Angew. Chem. Int. Ed., 2012, 51, 7812 7816.
Otten, LC, Fullam, E, Gibson, M.I. Molecular Biosystems, 2016, 12, 341-344.
Richards, S-J, Otten, LC, Gibson, MI, J. Mater. Chem. B., 2016, 10.1039/C5TB01994J
Richards, S-J., Fullam, E.C, Besra, G., Gibson, MI. J. Mater. Chem. B, 2014, 2, 1490-1498.
Jones, M. W., Otten, L, Richards, S-J, Lowery, R., Phillips D. J., Haddleton D. M., and Gibson MI. Chem. Sci., 2014, 5, 1611-1616.

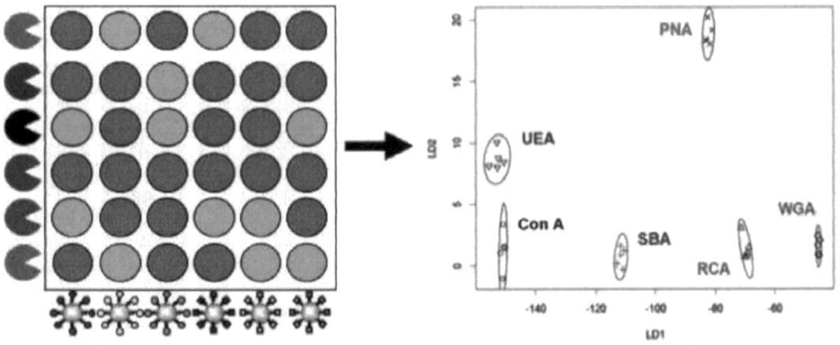

Multiplexed pathogen detection concept. Multiple glyconanoparticle channels, and their response to different lectins. Linear discriminants analysis (right) of the binding patterns.

Candace H. Haigler
North Carolina State University, Raleigh, NC, United States

Structure/function relationships in plant cellulose synthesis

The renewable cellulosic biomaterials made by plants are of premier importance in nature and industry. The strong crystalline cellulose microfibrils within plant cell walls are the product of a remarkable and dynamic cellular manufacturing process occurring at the interface of the cortical cytoplasm, plasma membrane, and apoplastic space. A soluble substrate (UDP-glucose) is polymerized into ß-1,4-glucan polymers by cellulose synthase (CESA) glycosyltransferases, which are grouped into a membrane-bound multimeric cellulose synthesis complex (CSC) that facilitates microfibril formation. Although no plant cellulose synthase structure has been solved empirically, the structural era of plant cellulose synthesis research began in 2013 with the generation of a high quality computational model of a CESA catalytic domain. This *de novo* folded predicted plant CESA structure was demonstrated to be useful by its congruence with the homologous part of the solved structure of the first cellulose synthase (from a bacterium) that was also published in 2013. The predicted plant CESA structure provided structural insights into the effects of known CESA missense mutations and at the same time highlighted structural differences between cellulose synthases that developed over evolutionary time. Ongoing work will be described to experimentally test plant CESA structure/function relationships, including empirical testing of CESA membrane topology, functional testing of a putative substrate binding loop in CESA, and the improvement and use of CESA models in combination with imaging to predict the number of CESAs within the CSC. This information improves our understanding of the regulation of plant cellulose synthesis and the size of the fundamental cellulose microfibril. Furthermore, it generates a platform for engineering CESA and the CSC to generate cellulose with improved properties for various industrial uses.

This work reflects the efforts of a multidisciplinary group supported as part of The Center for LignoCellulose Structure and Formation, an Energy Frontier Research Center funded by the U.S. Department of Energy, Office of Science, Basic Energy Sciences under Award # DE-SC0001090

Stephen Hanessian
Dept of Chem, Univ De Montreal, Montreal, QC, Canada

The challenging role of carbohydrates at the interface of chemistry and biology

Over the past century, the field of carbohydrate chemistry has evolved to become one of the focal points of many processes that define the complex nature of the structure to function to property continuum in biology. Among the many fascinating features of bioactive carbohydrate molecules or natural products containing sugar units is their interaction with physiologically relevant enzymes, receptors and macromolecules such as DNA and RNA. Aside from conformational, stereochemical and functional requirements, the anomeric configuration of glycosidically attached carbohydrates is a critical component in biological recognition and function. It follows that conceptually novel methods toward the stereocontrolled synthesis of glycosides and bioconjugates in general are worthy research areas of exploration, as will be highlighted in this tribute to Derek Horton, a giant in the field of carbohydrate chemistry.

Laura Hartmann
Institute for Organic Chemistry and Macromolecular Chemistry, Heinrich-Heine University, Düsseldorf, Germany

Precision glycopolymers as new tools to probe the multivalent binding of glycomimetics

Glycopolymers have been recognized as an important class of glycomimetics mimicking the multivalent presentation of single carbohydrate ligands on a macromolecular scaffold. They have been studied in terms of their role as model systems to investigate the effects and mechanisms of multivalency in ligand-receptor interactions.

Previously, our working group has demonstrated the synthesis of a new class of glycopolymers, the so-called precision glycomacromolecules. In contrast to polydisperse glycopolymers derived via classical polymer synthesis, precision glycomacromolecules are monodisperse and sequence-defined – much like their natural analogues, the glycopeptide and proteins. Thus they allow for detailed structure-activity studies on the binding of glycomacromolecules to protein receptors and help to promote the application of glycomimetics in biotechnology and biomedicine.

This talk will present the solid phase synthesis of precision glycomacromolecules from tailor-made building blocks using Fmoc-chemistry. Examples of the synthesis of monodisperse homo-and heteromultivalent systems and their evaluation with model lectin Concanavalin A (ConA) using different techniques such as SPR, ITC, fluorescence spectroscopy and STDNMR will be described. Our first findings demonstrate that the number and spacing of carbohydrate ligands on the macromolecular scaffold influence the multivalent binding to lectin receptor ConA.

Albert J. Heck
Univ Utrecht, Biomolecular Mass Spectrometry, Utrecht, Netherlands

A mass spectrometry view into diverse aspects of glycobiology

Around for more than a century the analytical technique of mass spectrometry is blooming more than ever, and applied in nearly all aspects of the natural and life sciences. In the last two decades mass spectrometry has become routine for the high-throughput analysis of peptides and glycans, and to a lesser extent glycopeptides. However, also intact proteins and even complete protein complexes can nowadays be analyzed. In this lecture, I will describe the emerging role of mass spectrometry with its different technical facets in molecular and structural biology, focusing especially on the analysis of intact glycoproteins. Moreover, I will describe how we use native mass spectrometry to study dynamic protein assemblies, especially those involved in complement activation.

Recent developments in mass spectrometry technology have allowed us to analyze intact native glycoproteins and protein complexes using Q-ToF and Orbitrap mass analyzers with very high sensitivity and mass resolving power, enabling us to profile the quality and biosimilarity of protein biotherapeutics, in their native state without requiring much sample preparation. In detail, I will demonstrate how native mass spectrometry can be combined with middle-down proteomics to profile complex structures of various glycoproteins, focusing on mAbs, plasma derived factor P, and Erythropoietin.

Furthermore, I will describe how native mass spectrometry has evolved into a tool that can provide unique structural and functional information about the assembly and topology of protein assemblies involved in complement activation, analyzing complexes of more than 40 subunits, when associated together exhibiting molecular weights over 2 million Da in mass. These data have provided new insight in how the complement pathway may be specifically activated by hexameric IgGs.

Visualization of the protein assembly involved in Complement activation by cryo-EM

Linda C. Hsieh-Wilson

Chemistry and Chemical Engineering, California Institute of Technology, Pasadena, CA, U.S.

Synthetic probes for understanding glycosaminoglycan recognition and signaling in the brain

Glycosaminoglycans (GAGs) comprise a large family of sulfated polysaccharides that regulate diverse biological events such as embryonic development, viral invasion, cancer metastasis, and spinal cord injury. Assembled from repeating disaccharide subunits, GAGs exhibit subtle variations in stereochemistry, chain length, and patterns of sulfation. This structural diversity is thought to enable the generation of a large number of protein-binding motifs. We will describe the synergistic application of organic chemistry and neurobiology to understand how specific GAG structures interact with protein receptors in the brain. In addition, we will discuss recent work on GAG-based polymer mimetics, which have enabled the first explorations into the importance of macromolecular structure on GAG function. These mimetics may also provide agents for modulating specific GAG-mediated processes *in vivo*. By combining synthetic organic chemistry, polymer chemistry, computational chemistry, and molecular and cellular neurobiology, our studies provide mechanistic insights into how GAGs regulate proteins and signaling events that underlie key processes such as neurite outgrowth, axon regeneration, and neural circuit formation.

Osamu Kanie
Institute of Advanced Biosciences, Tokai University, Hiratsuka, Kanagawa, Japan
Space, time and glycan processing: Synthetic and analytical approach

Synthesis of glycoconjugates is achieved in Golgi apparatus. However, its process is not fully understood, although the key players including glycosyltransferases and sugar nucleotide transporters have been characterized. Golgi apparatus is relatively large organella, which occupies an area close to that of nuclei, and is somewhat elusive provably because of its vesicular characteristics. Inside this, glycan structures are formed in a time dependent manner, however, time-resolved investigation of glycan synthesis is seldom challenged, and is considered to be the next issue to be addressed.

To start with, we decided to "visualize" this glycan processing events using fluorescently labeled lactosyl ceramide (LacCer) analogues utilizing various techniques. Here, the visualization includes 1) cellular localization analysis using confocal laser scanning microscopy, 2) structural analysis of glycans by using MS, and 3) analysis of mobility of the probe at plasma membrane and the Golgi region based on fluorescence recovery after photobleaching. The important aspects of this investigation are the time-resolved analyses. Pioneering investigation of BODIPY tagged Lac sphingosine (Sph) has been carried out by Pagano *et al*. Based on their findings, we set our visualization program to understand the glycan processing at Golgi. The attempt to enhance spatio resolution based on pulse-chase analysis made the structural analysis more difficult because of the limited amount of the probe introduced into the cultured cells. We further investigated speed of probe movement at plasma membrane and at Golgi region, and found that dramatic velocity difference possibly reflecting molecular diffusion and cisterna maturation. LacCer could be delivered into cis-Golgi, however, there is no known general method to send an external glycan into cellular organella. We have also shown that an artificial glycoprotein can be delivered into ER through cis-Golgi by means of retrograde transport.

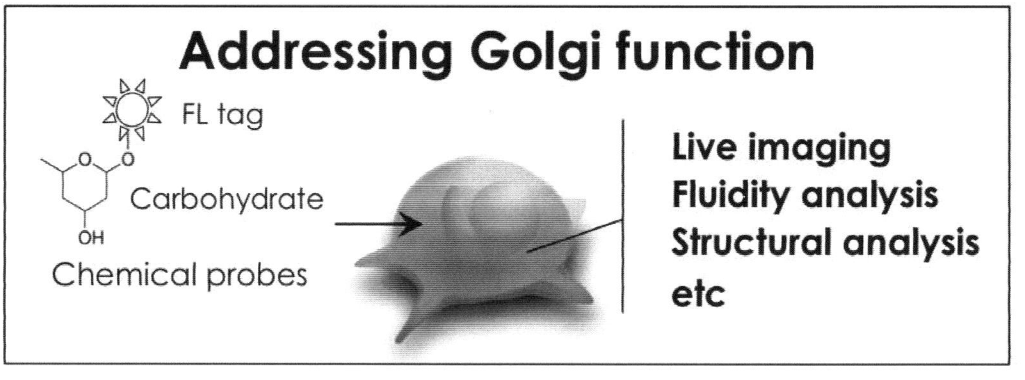

Laura L. Kiessling
Dept of Chemistry Biochemistry, University of Wisconsin Madison, Madison, WI, United States
Lectins as microbial detectors

A microbial community lives in and on us; we therefore need mechanisms to distinguish our cells from those of microbes. The glycans on the surface of mammalian cells can be assembled from about 35 building blocks while microbial glycans require over 700 different carbohydrate components. Lectins are uniquely poised to detect these differences and thereby distinguish microbes from animal cells. We envision that by understanding the binding preferences of human lectins, they can be exploited to combat infectious disease and to regulate microbiome composition to promote human health. Our studies of one class of lectins, the X-type lectins, will be described. These X-type lectins are found in chordates, and human

X-type lectins, intelectin-1 and intelectin-2, are produced at mucosal barriers and mutations in these proteins are associated with diabetes, cancer, and inflammatory/immune diseases.

Yanmei Li
Department of Chemistry, Tsinghua University, Beijing, China

Glycopeptide vaccines for cancer immunotherapy

Tumor cells aberrantly express carbohydrates and peptides, which can be considered as promising antigens for the development of tumor vaccines. Many efforts have been taken to increase the immunogenicity of antigens. In our works, we chose MUC1 glycopeptides as the tumor-associated antigens and developed several kinds of glycopeptide vaccines. Firstly, MUC1 glycopeptide was conjugated with Bovine Serum Albumin (BSA) to improve the immunogenicity [1]. However, BSA is highly immunogenic and can suppress immune reaction, the improved vaccines containing T-helper cell epitopes were developed [2]. To our knowledge, Toll-like receptor 2 ligand Pam_3CSK_4 is a potent immunostimulant. Therefore, we developed a three component chemically synthetic vaccine containing Pam_3CSK_4 and T-helper cell epitope
[3]. We also developed a noncovalent strategy to overcome the difficulty of covalent linkage strategy. [4] To further improve the efficiency of vaccines, cluster effect based on the dendritic lysine was used in the vaccines [5]. To avoid the side effect of the extra adjuvant, a self-assembly adjuvant-free vaccine was developed [6]. Besides, to utilize the size effect of the vaccines, we developed a new strategy for constructing an antitumor vaccine of glycopeptide nanoconjugates based on the multilayer self-assembly through the electrostatic interaction. All of these vaccines can stimulate the immune system to elicit high level of antibodies against MUC1 glycopeptides.

References

[1] Cai, H.; Huang, Z. H.; Shi, L.; Sun, Z. Y.; Zhao, Y. F.; Kunz, H.; Li, Y. M. Angew. Chem. Int. Ed. 2012, 51, 1719-1723.
[2] Cai, H.; Chen, M. S.; Sun, Z. Y.; Zhao, Y. F.; Kunz, H.; Li, Y. M. Angew. Chem. Int. Ed. 2013, 52, 6106-6110.
[3] Cai, H.; Sun, Z. Y.; Huang, Z. H.; Shi, L.; Zhao, Y. F.; Kunz, H.; Li, Y. M. Chem.Eur. J. 2013, 19, 1962-1970.
[4] Gao, Y.; Sun, Z. Y.; Huang, Z. H.; Chen, P. G.; Chen. Y. X.; Zhao, Y. F.; Li, Y. M. Chem.Eur. J. 2014, 20, 13541-13546.
[5] Cai, H.; Sun, Z. Y.; Chen, M. S.; Zhao, Y. F.; Kunz, H.; Li, Y. M. Angew. Chem. Int. Ed. 2014, 53, 1699-1703.
[6] Huang, Z. H.; Shi, L.; Ma, J. W.; Sun, Z. Y.; Cai, H.; Chen, Y. X.; Zhao, Y. F.; Li, Y. M. J. Am. Chem. Soc. 2012, 134, 8730-8733.

Hua Chai, Kim Le Mai Hoang, **Xue-Wei Liu**
Division of Chemistry and Biological Chemistry, Nanyang Technological University, Singapore, Singapore

Quick assembly of glycosides and glycoproteins: General tactics and new glycosylation methods

Glycopeptides and glycoproteins play vital roles in a wide array of biological activities including cell-cell adhesion, cell differentiation, cell growth and tumour metastasis. In addition, a host of major illnesses such as autoimmune diseases, infectious diseases and cancer are caused by aberrant protein glycosylation. Consequently, the development of new glycopeptides-based vaccines, diagnostics and therapeutics has gained significant interests, stimulating the study of glycopeptide synthesis. Here, a practical approach towards

N-glycopeptide synthesis using an auxiliary-mediated dual native chemical ligation (NCL) is presented. The first NCL connects an N-linked glycosyl auxiliary to the thioester side chain of an N-terminal aspartate oligopeptide. This intermediate undergoes a second NCL with a C-terminal thioester peptide. Mild deprotection provides the desired N-glycopeptide. The flexibility and systematic control of our method clearly have significant advantages over preceding ones and present a major step forward in glycoscience. Taking into consideration of the previous approaches, our method offers an attractive and practical procedure to N-glycopeptide that does not require cysteine residue, allowing for straightforward glycosylation and ligation at aspartic acid terminal. The auxiliary-linked glycosyl donor can be obtained from readily available glycosyl azide. Our method can leverage modern techniques such as intein-mediated ligation or recombinant expression to gain access to peptides of virtually any length and order.

Kevin J. Naidoo
1. Scientific Computing Research Unit, University of Cape Town, Rondebosch, 7701, South Africa; 2. Dept. of Chemistry, University of Cape Town, Rondebosch, 7701, South Africa.

Computing in cancer from glycobioinformatics, analytics based Glyco gene signature discovery to the simulation of glycosyltransferase mechanisms

Complex carbohydrates play a central role in cellular communication and in disease development. O-and N-glycans bonded to proteins and lipids on the cell surface make up the lion share of known cancer biomarkers. We have built an open access Glycome Analytics Platform (GAP) that connects multiple online glycan resources that can be used to design reproducible data analytic experiments for obtaining, generating and analyzing cell glycomes. The GAP links databases such as KEGG and the CFG to a suite of tools and utilities and is integrated into the analytic functionality of the Galaxy bioinformatics platform. However, recognizing that tumour glycome data are sparse we deployed machine learning classification tools onto glycosyltransferase (GT) gene expression profiles of tumours from six cancers sourced from The Cancer Genome Project (TCGA). We found that the expression profiles of 210 GT genes taken from 1893 cancer patient samples in TCGA microarray data are able to classify six cancers; breast, ovarian, glioblastoma, kidney, colon and lung. We found that the GT gene expression profiles can be used to develop cancer classifiers and propose subtypes. The GT gene regulatory patterns are observed to match known GT targets for breast cancer such as ST3Gal-I and ST6Gal-I. In downstream computer experiments we use our Free Energy from Adaptive Reaction Coordinate Forces (FEARCF) method to map out detailed sialyltransferase reaction mechanisms. In a tour of our glycobiology dry lab I will describe the value of open source collaborative glycomics platforms, illustrate cancer biomarker discovery from big data analytics and showcase specialized glycan software tools that are central to the design of inhibitors of GT targets.

Vadim Krylov, Dmitry Argunov, Dmitry Vinnitskiy, Alexey Gerbst, Nadezhda Ustyuzhanina, **Nikolay Nifantiev**
Glycoconjugates, N.D. Zelinsky Institute of organic chemistry, RAS, Moscow, Russian Federation

Pyranosideintofuranoside rearrangement: New reaction in carbohydrate chemistry and its application in oligosaccharide synthesis

Great interest in natural furanoside-containing compounds has challenged the development of preparative methods for their synthesis. Herein we report a novel reaction in carbohydrate chemistry, namely a pyranoside-*into*-furanoside (PIF) rearrangement permitting the transformation of selectively O-substituted pyranosides into the corresponding furanosides [1,2]. The discovered process, which has no analogy in organic chemistry, includes acid-promoted per-O-sulfation of pyranoside which is accompanied by ring rearrangement into furanoside one followed by solvolytic O-desulfation. This process was shown to be a very useful tool for the synthesis of furanoside-containing complex oligosaccharides, which was demonstrated by synthesizing oligosaccharide related to O-chain of the lipopolysaccharides from Gram negative bacterium *Klebsiella pneumoniae* [1], diheteroglycan from Gram positive cocci *Enterococcus faecalis* [1,3], galactomannan of the fungus (mold) *Aspergillus fumigatus* [4], and the fucoidan from brown seaweed *Chordaria flagelliformis* [5].

This work was supported by RSF grant 14-23-00199.

[1]. V.B. Krylov, D.A. Argunov, D.Z. Vinnitskiy, S.A. Verkhnyatskaya, A.G. Gerbst, N.E. Ustyuzhanina, A.S. Dmitrenok, J. Huebner, O. Holst, H.C. Siebert, N.E. Nifantiev, *Chem. Eur. J.*, 2014, 20, 16516.
[2]. V.B. Krylov, D.A. Argunov, D.Z. Vinnitskiy, A.G. Gerbst, N.E. Ustyuzhanina, A.S. Dmitrenok, N.E. Nifantiev, *Synlett*, 2016, DOI: 10.1055/s-0035-1561595.
[3]. V. B. Krylov, A. G. Gerbst, D. A. Argunov, A. S. Dmitrenok, A. S. Shashkov, Z. Kaczynski, J. Huebner, O. Holst, and N. E. Nifantiev, *Chem. Eur. J.*, 2015, 21, 1749.
[4]. D.A. Argunov, V.B. Krylov, N.E. Nifantiev, *Org. Biomol. Chem.*, 2015, 13, 3255.
[5]. D.Z. Vinnitskiy, V.B. Krylov, N.E. Ustyuzhanina, A.S. Dmitrenok, N.E. Nifantiev, *Org. Biomol. Chem.*, 2016, 14, 598.

Mark Nitz
Chemistry, University of Toronto, Toronto, ON, Canada

The production and inhibition of biofilm related exopolysaccharides

Biofilms are antibiotic resistant communities of bacteria that adhere to biotic and abiotic surfaces. The formation of bacterial biofilms requires an extracellular matrix to facilitate adherence of bacteria to the surface they colonize. A wide variety of medically important bacterial strains, including *S. epidermidis*, S. aureus, *E. coli*, *B. pertussis*, and *Y. pestis* generate similar partially de-N-acetylated β(1-6-)-N-acetylglucosamine (PNAG) homopolymers as a key biofilm matrix exopolysaccharides, making their biosynthesis an important target for anti-infectives. In these bacterial strains PNAG de-N-acetylation is essential for polysaccharide export and surface attachment. The biosynthesis of PNAG involves a series of steps where the ultimate or penultimate step is de-N-acetylation. This amide hydrolysis is carried out by enzymes found in the carbohydrate esterase family 4. This seminar will focus on the de-N-acetylases and synthase complex necessary to produce PNAG. To date the de-N-acetylases have shown low catalytic activity *in vitro*, surprising substrate specificity, and unusual metal dependence in relation to other characterized carbohydrate de-N-acetylases. A series of inhibitors have been synthesized based on phosphonamidates, phosphinates, phosphonates and trifluoromethyl ketones to target these enzymes. Recent work on the synthase mechanism will also be discussed.

Markus Pauly
Plant Cell Biology and Biotechnology, Heinrich Heine University Düsseldorf, Düsseldorf, Germany

Synthetic biology of plant cell wall polysaccharides

All plant cells are encased in a complex composite, the plant cell wall. The structure of this wall has evolved to benefit the sessile lifestyles of plants allowing their survival for up to several millennia. Plant cell walls consist of several polysaccharide networks; a complex aggregate of microcrystalline cellulose microfibrils, crosslinking water soluble hemicelluloses, and acidic pectic polysaccharides. Major advances have been made in identifying the proteins that are involved in the synthesis of these polysaccharides. Genetic manipulation of this machinery allows for the synthesis of novel polysaccharides in the plant AND their functional assessment in the life cycle of the plant. Studying the biosynthesis of the wall polysaccharides will aid us in achieving our ultimate goal, to one day create tailored plant polysaccharide structures in a microbe or even cell free.

Jennifer A. Prescher
Chemistry Department, University of California Irvine, Irvine, CA, United States

Expanding the imaging toolbox

The chemical reporter strategy has emerged as a powerful method to equip glycans and other scaffolds with visual probes. This strategy involves the metabolic introduction of unique chemical handles (i.e., "chemical reporters") into target biomolecules. The reporter groups can then be specifically detected utilizing highly selective (i.e., bioorthogonal) reactions. While dozens of chemical reporters and bio-orthogonal reactions have been reported in recent years, the majority suffer two significant drawbacks: (1) the reagents are too large or unstable for tagging biomolecules in live cells and (2) many of the most popular reagents cross react with one another and cannot be used concurrently. This has stymied our ability to visualize collections of biomolecules. To address these limitations, we are identifying new reactions and combinations of compatible reactions that can be used concurrently in live cells. Over the past few years, we have added three new reagents to the bio-orthogonal toolkit – cyclopropenes, cyclopropenones, and triazines. These motifs are remarkably biocompatible and can be tuned to access diverse classes of reactions. Many of the chemistries can also be used in tandem, enabling multicomponent imaging studies. The development of the bio-orthogonal reagents, along with their application to tracking secreted metabolites, will be discussed.

Orlando J. Rojas
Forest Products Technology, Aalto University, Espoo, Helsinki, Finland

Interfacial behavior of carbohydrates: from amphiphilic sugars to cellulose nanocrystals

The effects of nonionic sugar surfactant adsorption at the air/liquid interface were studied by surface laser light scattering by capillary waves. The viscoelastic properties (surface elasticity and surface viscosity) of monolayers formed by octyl and 2-ethylhexyl glucoside isomeric surfactants were quantified at submicellar concentrations and a diffusional relaxation model was found to describe the observed trends. The interfacial diffusion coefficients that resulted from fitting the diffusional relaxation model to surface elasticity values reflect the molecular dynamics of the sub-phase near the interface. Our results demonstrate important differences in molecular packing at the air-water interface. In contrast, a much slower dynamics, is found in the case of assemblies of cellulose forming amphiphilic crystals (cellulose nanocrystals, CNC). In such cases adsorption at the air-water interface is only observable in the presence of insoluble cationic surfactants. However, when used to stabilize oil-water systems, CNC are effective in forming structured networks. Taking advantage of this property and also their amphiphilicity, novel Pickering emulsions are easily obtained. This is relevant to many fields, including pharma and food and material synthesis (fibers, films or particles). The main principles used in these novel approximations for the valorization of biomass, especially in the context of the behavior of carbohydrate-based amphiphiles, will be discussed in light of the enormous potential and value of formulation of related multiphase systems.

Frank Schroeder
1. Boyce Thompson Institute, Ithaca, NY, United States; 2. Chemistry and Chemical Biology, Cornell University, Ithaca, NY, United States

Two 3,6-dideoxyhexoses serve as scaffolds for a combinatorial library of signaling molecules in *C.elegans*

The simple nematode *Caenorhabditis elegans* has become one of the most important model organisms for biomedical research because many of its physiological pathways show strong homology to corresponding pathways in humans. We aim to complement the highly developed genomics and proteomics of *C. elegans* with a comprehensive structural and functional characterization of its metabolome, which has been explored to only a very limited extent.

We found that worms are amazingly skilled chemists: using simple building blocks from conserved primary metabolism and a strategy of modular assembly, *C. elegans* and other nematode species create complex molecular architectures to regulate almost every aspect of their development and behavior. Most of the identified compounds are based on two 3,6-dideoxyshexoses, L-ascarylose and L-paratose. Whereas L-ascarylose and D-paratose are known from bacteria, L-paratose had not previously been identified from natural sources. In nematodes, L-ascarylose and L-paratose serve as scaffolds for attachment of moieties from amino acid, carbohydrate, neurotransmitter, lipid, and nucleoside metabolism, including an unusual xylopyranose-based adenosine derivative. The resulting signaling molecules can be active at femtomolar concentrations, such that encountering just a few molecules per minute is sufficient for worms to respond. Identification and biosyntheses of these compounds continue to reveal new aspects of primary metabolic pathways, from carbohydrate and lipid metabolism to endocannabiod biosynthesis and nucleoside catabolism. The discovery of new types of modular, primary metabolism-derived signaling molecules in *C. elegans* provides a strong incentive for a comprehensive reanalysis of metabolism in higher animals, including humans.

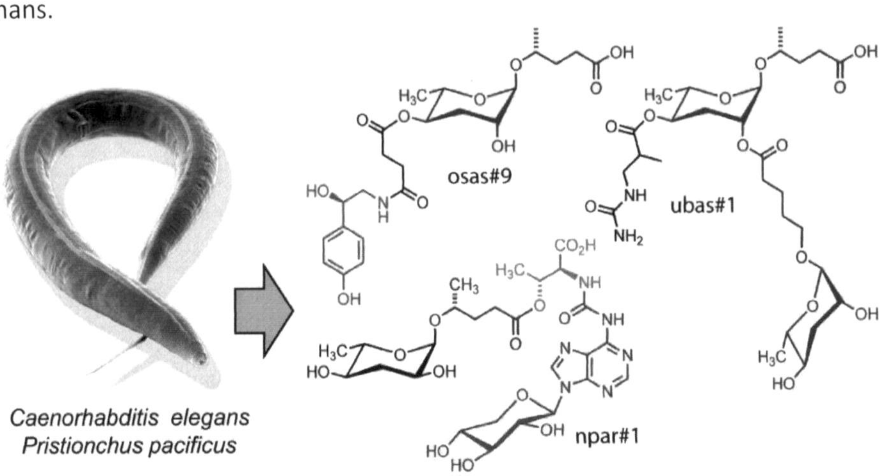

Nematode model organisms have revealed a combinatorial library of signaling molecules derived from two 3,6-dideoxyhexoses, L-ascarylose and L-paratose.

Peter H. Seeberger

Dept of Biomolecular Systems, Max Planck Inst of Colloids Interfaces, Berlin, Germany

Carbohydrate vaccine development enabled by automated glycan assembly

Pure glycans are key to enable biochemical, biophysical and immunological studies aimed at understanding the role of carbohydrates. Described is the development of a fully integrated platform for automated glycan assembly (AGA) based on solid phase oligosaccharide sythesis[1] and carbohydrate arrays to address biological problems. Particular emphasis in this lecture will be placed on the new automated synthesis platform[2] that has been commercialized. [3,4] Access to defined polysaccharides as long as 50mers enables now biological investigations.[5] Quality control of synthetic glycans can now be guaranteed using ion mobility mass spectrometry to very low levels. [6]

Carbohydrate arrays are used as diagnostics and in support of vaccine programs that are based on conjugates with synthetic oligosaccharides to screen blood sera.[7] Case studies of specific vaccines will provide an appreciation for the approach that is now advancing candidates toward clinical testing. [8,9] Fully synthetic vaccine candidates exploit iNKT cells for to induce a robust and protective immune response.[10]

References
1. Plante, O.J.; Palmacci, E.R.; Seeberger, P.H.; Science 2001, 291, 1523.
2. Castagner, B.; Kröck, L.; Esposito, D.; Wang, C.C.; Bindschädler, P.; Seeberger, P.H.; Chem. Sci. 2012, 3, 1617.
3. www.glycouniverse.de
4. Seeberger, P.H.; *Acc. Chem. Res.* 2015, 48, 1450.
5. Kottari, N.; Schuhmacher, F.; Seeberger, P.H.; manuscript in preparation.
6. Hofmann, J.; Hahm, H.S.; Seeberger, P.H.; Pagel, K.; *Nature* 2015, 526, 241.
7. Kamena, F.; Tamborrini, M.; Liu, X.; Kwon, Y.U.; Thompson, F.; Pluschke, G.; Seeberger; P.H.; *Nature Chem. Bio.*, 2008, 4, 238.
8. Schofield, L.; Hewitt, M.C.; Evans, K.; Siomos, M.A.; Seeberger, P.H.; *Nature,* 2002, 418, 785.
9. Martin, C.E.; Broecker, F.; Oberli, M.A.; Komor, J.; Mattner, J.; Anish, C.; Seeberger, P.H.; *J. Am. Chem. Soc.* 2013, 135, 9713.
10. Cavallari, M.; Stallforth, P.; Kalinichenko, A.; Rathwell, D.; Gronewold, T.M.A.; Adibekian, A.; Mori, L.; Landmann, R.; Seeberger, P.H.; De Libero, G.; *Nature Chem. Bio.* 2014, 10, 950.

Wenhui Zhang[3], MiKyung Yoon[1], Ian Carmichael[2], **Anthony S. Serianni** [1]
1. Chemistry and Biochemistry, University of Notre Dame, Notre Dame, IN, USA; 2. Radiation Lab, University of Notre Dame, Notre Dame, IN, United States; 3. Univesity of Notre Dame, Granger, IN, United States

Synthesis, isotopic labeling, and structural studies of high-mannose oligosaccharides

This laboratory has been developing methods to introduce stable isotopes such as ^{13}C, ^{2}H and ^{15}N into saccharides for > 30 years. Numerous studies have demonstrated that labeling promotes a deeper understanding of saccharide structure and reactivity. For example, the introduction of ^{13}C enables measurements of ^{13}C-^{13}C spin-couplings, and improves measurements of ^{13}C-^{1}H spin-couplings, which when used collectively with J_{HH} values lead to firmer conformational assignments, especially in flexible molecules. The value of labeling increases with saccharide size; NMR analyses are simplified through, for example, (1) the incorporation of one or more perdeuterated residues to reduce signal density, and (2) the use of ^{13}C nuclei to edit complex datasets. With these objectives in mind, effort has been devoted to defining the structural dependencies of the abundant ^{1}J, ^{2}J and ^{3}J values involving ^{13}C, ^{1}H and ^{15}N in saccharides. While J-couplings are local structure constraints, they are valuable parameters to assist in, and validate, structural interpretations of longer range constraints such as RDCs. This talk will summarize recent work to prepare high-mannose oligosaccharides in high purity and in milligram quantities containing multiple ^{13}C labels. Chemical and enzymic methods to prepare N-glycans in general, optimal isotopic labeling strategies, and the integration of DFT calculations with experimental NMR measurements to assign oligosaccharide structure in solution, will be treated. Emerging measurements of ^{13}C-^{13}C spin-couplings in defined molecular geometries by solid state NMR and x-ray crystallography will also be discussed.

[This work was supported by National Science Foundation CHE 1402744 and National Institutes of Health SBIR HHSN261201500020C.]

Daan van Aalten
1. School of Life Sciences, University of Dundee, Dundee, United Kingdom

Mechanisms and inhibitors of O-GlcNAc signalling

The reversible modification of intracellular proteins with O-linked N-acetylglucosamine (O-GlcNAc) is essential for embryonic development in metazoa. Inhibitors of O-GlcNAc transferase (OGT), the enzyme that catalyses this modification, are valuable tools to study the cell biology of protein O-Glc-N-Acylation. By determining OGT catalytic mechanism, revealing unusual participation of the pyrophosphate moiety, and determining OGT substrate specificity by screening random peptide libraries, we have identified suitable components for the construction of bisubstrate inhibitors. We report OGT bisubstrate linked inhibitors (goblins) in which the acceptor serine in the peptide VTPVSTA is covalently linked to UDP, eliminating the GlcNAc pyranoside ring. Aided by structural biology and a novel fluorescence polarimetry assay, we have obtained goblins that bind the enzyme with single digit micromolar affinity, inhibiting glycosyltransfer onto protein and peptide substrates. I will report our latest results in optimising their cell penetrance and effects on Drosophila embryonic development.

Hans H. Wandall
Department of Cellular and Molecular Medicine, University of Copenhagen, Denmark
Molecular dissection of glycan function by simple cells and simplified tissues

Our molecular understanding of how glycans influence cellular functions is limited. This is in part due to the complexity and technical difficulties associated with the analysis of glycans. Precise genome editing techniques now enable us to generate glycoengineered cell lines suitable for several purposes, including biopharmaceutical production, genetic dissection of glycan functions, formation of organotypic tissue models, and arraying platforms for host pathogen interactions. As the first proof-of-concept for our discovery strategy, we have used genetic engineering to deconstruct N-glycosylation in CHO cells and developed a design matrix that facilitates the generation of cells with desired glycosylation. This engineering approach will aid the production of glycoproteins with defined and homogenous glycosylation and allow genetic dissection of glycan functions in many different cells. In addition, we have pioneered a next generation approach using organotypic tissue models in combination with sophisticated mass spectrometry to decipher glycan functions. The tissue model has provided the first evidence that aberrant glycosylation in cancer directly induces oncogenic features. Moreover, the model has demonstrated that glycosylation of the envelope virus Herpes Simplex Virus 1 is essential for viral propagation. Future studies will aim to use step-by-step genetic deconstruction of glycosylation capacities and provide comprehensive cell and tissue models for discovery and dissection of structure-function relationships of glycans.

Chris Whitfield
Molecular Cellular Biology, University of Guelph, Guelph, ON, Canada
The structure and functions of bacterial β-Kdo glycosyltransferases

Kdo (3-deoxy-D-*manno*-oct-2-ulosonic acid) is an 8-carbon sugar mostly confined to bacteria. α-Kdo is a component of all bacterial lipopolysaccharides, where it links the hydrophobic lipid A moiety to the rest of the molecule. In contrast, β-Kdo is relatively rare. It is a component of glycan repeat units in a small selection of lipopolysaccharide O-antigenic polysaccharides and capsular polysaccharides. However, it is also present in a recently discovered conserved glycolipid acceptor for a class of capsular polysaccharides (CPS) produced by several Gram negative pathogens. The poly-β-Kdo region of this glycolipid links variable capsular polysaccharide structures to a terminal (lyso)phosphatidylglycerol residue. Synthesis of the poly-β-Kdo linker requires two proteins with CMP-β-Kdo-dependent glycosyltransferase (β-Kdo GT) activity; a monofunctional β-Kdo GT and a polymerizing enzyme containing two β-Kdo GT domains with different linkage specificities. The assembly pathway has been resolved using a combination of natural and synthetic oligosaccharide acceptors. The catalytic mechanism of β-Kdo GTs is unknown. They share no significant sequence similarity with inverting α-Kdo GTs and were not identified as GTs in the CAZy database. We solved the structure of a prototype β-Kdo

GT from *Raoultella terrigena* and *Klebsiella pnuemoniae*. The overall structure resembles the GT-B fold and the CMP-binding site architecture is similar to some sialyltransferases. However, each of the Rossmann-like α/β domains has a reduced number of strands and the N-terminal α/β domain is displaced from its usual position in canonical GT-B proteins (including α-Kdo GTs). An additional α-helical domain is positioned at the junction of the α/β domains and it shows little resemblance to any known structure. These studies provide new insight into the structure and function of a new family of GT enzymes that includes candidates for developing antivirulence agents against encapsulated pathogens.

The Technical Program

Sheila Murphy, *Program Chair*

SUNDAY MORNING, July 17

RESEARCH DIRECTIONS IN GLYCOSCIENCE

New Orleans Marriott, Galerie 3

A. D. French, *Organizer*
G. Boons, K. Kato, P. Marino, S. Perez, *Presiding*

10:00 Introductory Remarks

10:05 Panel Discussion

SUNDAY AFTERNOON, July 17

OPENING CEREMONY

New Orleans Marriott, Acadia Room

Welcome Addresses

S. M. Murphy, *Organizer*
A. D. French, *Presiding*

4:00 Alfred D. French (Symposium Chair)

4:05 Pamela Marino (USA National Institutes of Health)

4:10 N. Jayaraman (President, International Carbohydrate Organization)

4:15 J.F.G. Vliegenthart (Representative for the International Union of Pure and Applied Chemistry)

4:20 Meeting Overview. **Alfred D. French**

SUNDAY AFTERNOON, July 17

AWARD SESSION

New Orleans Marriott, Acadia Room

Roy L. Whistler International Award in Carbohydrate Chemistry 2016

S. M. Murphy, *Organizer*
N. Jayaraman, *Presiding*

4:30 1. Sugars and proteins. **Benjamin G. Davis**

5:15 Intermission – Preservation Hall

AWARD SESSION

New Orleans Marriott, Acadia Room

ICO Young Researcher Award 2016
Financially supported by TCI – Tokyo Chemical Industry Co., Ltd.

S. M. Murphy, *Organizer*
J. F. Vliegenthart, *Presiding*

5:45 2. Marking the mycomembrane: Chemical tools for studying and targeting glycolipids of the mycobacterial outer membrane. **Benjamin M. Swarts**

PLENARY SESSION
Plenary Lecture (1)

New Orleans Marriott, Acadia Room

S. M. Murphy, *Organizer*
J. A. Prescher, *Presiding*

6:15 3. Glycocalyx engineering toward probing cancer glycome evolution. **Carolyn R. Bertozzi**

MONDAY MORNING, July 18

PLENARY SESSION
Plenary Lecture (2)

New Orleans Marriott, Acadia Room
S. M. Murphy, *Organizer*; Anne Imberty, *Presiding*

8:30 4. Sialic acids as determinants of self. **James C. Paulson**, S. Duan, M.S. Macauley, R. McBride, C. Nycholat, L. Pang, A. Srivastava, R. de Vries

PLENARY SESSION
Plenary Lecture (3)

New Orleans Marriott, Acadia Room
S. M. Murphy, *Organizer*; Beat Ernst, *Presiding*

9:15 5. GalNAc-conjugated oligonucleotides as a new paradigm in RNAi therapeutics for human diseases. **Muthiah Manoharan**

10:00 Intermission – Preservation Hall

CARBOHYDRATE STRUCTURE, ANALYSIS & COMPUTATION
Conformation and Interactions (1)

New Orleans Marriott, Galerie 3
A. D. French, K. W. Moremen, R. J. Woods, *Organizers*; Serge Perez, *Presiding*

10:30 6. Keynote Lecture: Mechanisms and inhibitors of O-GlcNAc signalling. **Daan van Aalten**

11:00 7. A role for underlying glycan structure in influenza binding: extending the species specificity paradigm. **Robert J. Woods**, O.C. Grant, J.A. Hadden, Y. Ji, H.M. Smith, W. Peng, R. de Vries, R. McBride, J.C. Paulson

11:20 8. Complex carbohydrate recognition by proteins: Fundamental insights from bacteriophage cell adhesion systems. N.K. Broeker, Y. Kang, U. Gohlke, D. Andres, S. Kunstmann, A. Schmidt, M. Santer, **Stefanie Barbirz**

11:40 9. Conformational analysis of synthetic glycosylphosphatidylinositol (GPI) fragments using paramagnetic-NMR techniques. **Monika Garg**, M. Santer, K.N. Kirschner, D. Varón Silva

12:00 10. Hydrogen bonding and conformational changes in sialyl Lewis-X. M. Battistel, H. Azurmendi, M. Frank, **Daron I. Freedberg**

MONDAY MORNING, July 18

GLYCOBIOCHEMISTRY & GLYCOBIOLOGY
Proteoglycans and Sulfated Glycoproteins
Cosponsored by the Carbohydrate Division of ACS

New Orleans Marriott, Bissonet
L. L. Kiessling, J. C. Paulson, *Organizers*; K. Godula, *Presiding*

10:30 11. Keynote Lecture: Synthetic probes for understanding glycosaminoglycan recognition and signaling in the brain. **Linda C. Hsieh-Wilson**

11:00 12. Identification of modifier genes that modulate cell surface P-selectin ligand expression. **HiromuTakematsu**, H. KItagawa, R. Kannagi, S. Oka

11:20 13. Human heparanase – structure and function. **Liang Wu**, J. Jiang, Y. Jin, H.S. Overkleeft, G. Davies

11:40 14. Uncovering the relationship between sulfation patterns and conformation of iduronic acid in heparan sulfate. **Po-Hung Hsieh**, D.F. Thieker, M. Guerrini, R.J. Woods, J. Liu

12:00 15. Synthesis of disaccharide standards for LC-MS/MS quantitation of heparan sulfate in urine from mucopolysaccharidosis patients. **Qi Qi He**, P.J. Trim, J.J. Hopwood, M.F. Snel, V. Ferro

---

NEW DEVELOPMENTS IN GLYCOSCIENCE
Carbohydrates in Health and Disease (1)

New Orleans Marriott, Galerie 2
A. D. French, M. A. Godshall, S. M. Murphy, J. R. Vercellotti, *Organizers*; S. S. van Leeuwen, *Presiding*

10:30 16. Overcoming the obstacles associated with cryopreservation to enable modern cellular therapies. J. Briard, J. Poisson, T. Turner, S. Jahan, J. Acker, N. Pineault, **Robert Ben**

11:00 17. Carbon-based glyco-nanomaterials: A smart tool for the selective aggregation of bacteria and advanced drug delivery. **Noureddine Khiar**, M. Assali, M. Pernía-Leal, J.J. Cid-Martín, V. Valdivia, I. Fernández, R. Wellinger

11:20 18. Determining the variability of monoclonal antibody glycosylation profiles: A NIST interlaboratory study. **M. Lorna A. De Leoz**, D.L. Duewer, S.E. Stein

11:40 19. Glycopolymeric materials for selective inhibition of bacterial toxins. **Matthew I. Gibson**

12:00 20. Novel multivalent platforms for glycan presentation. **Stephen Hill**, M. Galan, M. Berry, D. Benito-Alifonso

MONDAY MORNING, July 18

SYNTHESIS OF CARBOHYDRATES, GLYCOCONJUGATES & GLYCAN-BASED BIOMATERIALS
Automated Synthesis
Cosponsored by the Carbohydrate Division of ACS

New Orleans Marriott, Galerie 5, 6
M. Cloninger, A. Demchenko, *Organizers*; G. Boons, *Organizer, Presiding*

10:30 21. <u>Keynote Lecture</u>: Carbohydrate vaccine development enabled by automated glycan assembly. **Peter H. Seeberger**

11:00 22. Improvement of stereoselectivity in the electrochemical glycosylation with mixed electrolyte systems. **Toshiki Nokami**, K. Kitamura, Y. Isoda, N. Sasaki, T. Itoh

11:20 23. Application of nanoporous gold to supported synthesis of carbohydrates and their biomolecular recognition. **Keith J. Stine**, A. Demchenko, A. Alla, J.K. Bhattarai, D. Neupane, S. Pistorio, V. Mikhaylov

11:40 24. Glycosylated gold nanoparticle biosensors: Label-free and high-throughput evaluation of glycan/lectin interactions. **Sarah-Jane Richards**, L. Otten, E. Fullam, M.I. Gibson

12:00 25. Sugars with conformationally locked 6-hydroxyl groups. **Daniel B. Werz**

SYNTHESIS OF CARBOHYDRATES, GLYCOCONJUGATES & GLYCAN-BASED BIOMATERIALS
Synthesis of Glycans

New Orleans Marriott, Acadia Room
G. Boons, M. Cloninger, A. Demchenko, *Organizers*; V. Wittmann, *Presiding*

10:30 26. <u>Keynote Lecture</u>: Molecular dissection of glycan function by simple cells and simplified tissues. **Hans H. Wandall**

11:00 27. Development and use of de novo asymmetric approaches to oligosaccharide motifs for carbohydrate medicinal chemistry. **George A. O'Doherty**

11:20 388. Un-natural cyclic oligosaccharides with expanded glycosidic bond. **Narayanaswamy Jayaraman**, G. Daskhan, M.G. Krishna, G.K. Samanta

11:40 29. Modular synthesis of *N*-glycans and arrays for the hetero-ligand binding analysis of HIV antibodies. **Chung Y. Wu**, S.S. Shivatare, C. Wong

12:00 30. Combining enzymatic with total synthesis offers a new path to build biologics. **Ivan A. Gagarinov**, G. Boons

MONDAY AFTERNOON, July 18

CARBOHYDRATE STRUCTURE, ANALYSIS & COMPUTATION
Conformation and Interactions (2)

New Orleans Marriott, Galerie 3
A. D. French, K. W. Moremen, *Organizers*; R. J. Woods, *Organizer, Presiding*

2:00 31. Keynote Lecture: Synthesis, isotopic labeling, and structural studies of high-mannose oligosaccharides. W. Zhang, M. Yoon, I. Carmichael, **Anthony S. Serianni**

2:30 32. Retaining glycosyltransferases mechanism(s) analyzed by QM/MM calculations. F. Mendoza, H. Gómez, J. Lluch, **Laura Masgrau**

2:50 33. Toll-like receptor 4: A computational walk into the molecular recognition process and modulation. **Sonsoles Martin-Santamaria**, L. Pérez-Regidor, J. Guzman-Caldentey, J. Billod, J. Klett, A. Lacetera

3:10 34. Investigation of the conformation of carbohydrates in the gas and micro-hydrated phases. **Oliver Schwardt**, M. Kandziora, B. Wagner, B. Ernst, G. O'Connor, P. Çarçabal, M. Smieško

3:30 35. Ring inversion properties of hexopyranoses: From force field parametrization to studies on the interplay between ring shape and glycosidic linkage conformation. **Wojciech Plazinski**, A. Plazinska, M. Drach

3:50 Intermission, Exhibits, Posters – Preservation Hall

GLYCOBIOCHEMISTRY & GLYCOBIOLOGY
Glycans and Immunity

New Orleans Marriott, Bissonet
L. L. Kiessling, J. C. Paulson, *Organizers*; Xing Chen, *Presiding*

2:00 36. Keynote Lecture: Glycopeptide vaccines for cancer immunotherapy. **Yanmei Li**

2:30 37. Increasing carbohydrate antigen immunogenicity by targeting MGL2 receptors with bivalent PS A1 conjugates. **Peter R. Andreana**

2:50 38. Immune cell targeting with liposomal nanoparticles displaying high-affinity glycan analog ligands of Siglecs. **Corwin Nycholat**, S. Duan, R. McBride, K. Worrell, J.C. Paulson

3:10 39. Design, synthesis and evaluation of optimized *Quillaja* saponin vaccine adjuvants. **A. Fernandez Tejada**

3:30 40. Glycosylation in cancer: glycomic strategies for characterization of molecular functions and biomarker applications. **Celso Reis**

3:50 Intermission, Exhibits, Posters – Preservation Hall

MONDAY AFTERNOON, July 18

GLYCOBIOCHEMISTRY & Glycobiology
Host-Pathogen Interactions

New Orleans Marriott, Galerie 5, 6
L. L. Kiessling, J. C. Paulson, *Organizers*; F. Schroeder, *Presiding*

2:00 41. <u>Keynote Lecture</u>: Glycosylated nanoparticles for the detection of pathogens and toxins. **Matthew I. Gibson**

2:30 42. H7N9 and H6N1 influenza A virus hemagglutinins engineered to bind human type receptors reveal a novel layer of specificity beyond the **α**2-6 linkage of sialic acid. **Robert P. de Vries**, W. Peng, R. McBride, J.C. Paulson

2:50 43. Furanoside metabolism to target *Leishmania* parasites. **Pierre Lafite**, J. Ati, T. Rattier, S. Belaz, J. Gangneux, F. Robert-Gangneux, F.H. Routier, R. Daniellou

3:10 44. The structural characterization of a novel high-affinity sialic acid specific CBM40 from *Clostridium perfringens* and the engineering of its divalent form. **João P. Ribeiro**, W. Pau, C. Pifferi, O. Renaudet, A. Varrot, L.K. Mahal, A. Imberty

3:30 45. How bacteria invade the bladder. **Maximillian M. Sauer**, R.P. Jakob, J. Eras, S. Baday, D. Eris, G. Navarra, S. Bernèche, B. Ernst, T. Maier, R. Glockshuber

3:50 Intermission, Exhibits, Posters – Preservation Hall

MONDAY AFTERNOON, July 18

NEW DEVELOPMENTS IN GLYCOSCIENCE
Synthesis: Methods and Applications (1)

New Orleans Marriott, Galerie 2
A. D. French, M. A. Godshall, S. M. Murphy, J. R. Vercellotti, *Organizers*; P. H. Seeberger, *Presiding*

2:00 46. Synthesis of selectively acetylated 2-deoxy-D-glucoses, potential agents for treatment of brain tumors. **Izabela Fokt**, M. Cybulski, S. Skora, W. Priebe

2:30 47. *C*-Glucosyl flavonoid analogues with neuroprotective effects: rational design and synthesis of new CNS-targeted drug-like leads against Alzheimer's disease. **Ana Marta de Matos**, J. Cristóvão, C. Gomes, A.P. Rauter

2:50 48. Improved Fe (III)-catalyzed selective benzylation of carbohydrates by Ag_2O and TBAB. **Hai Dong**

3:10 49. Enzymatic production of potential solubilizing agents, cycloisomaltomegalosaccharide and α-1,3 branched cycloisomaltomegalosaccharide. **Kazumi Funane**, R. Suzuki, H. Ichinose, M. Takatsugu, D. Mizushima, K. Kimura, Z. Fujimoto, S. Suzuki, S. Kitamura, A. Kimura, H. Hara

3:30 50. Synthesis of a carbocyclic activity-based probe for sialidases. **Jorin Hoogenboom**, H. Zuilhof, T. Wennekes

3:50 Intermission, Exhibits, Posters – Preservation Hall

SYNTHESIS OF CARBOHYDRATES, GLYCOCONJUGATES & GLYCAN-BASED BIOMATERIALS
New Reactions in Carbohydrate Synthesis

New Orleans Marriott, Acadia Room
G. Boons, M. Cloninger, *Organizers*; A. Demchenko, *Organizer, Presiding*

2:00 51. Keynote Lecture: Pyranoside-*into*-furanoside rearrangement: New reaction in carbohydrate chemistry and its application in oligosaccharide synthesis. V. Krylov, D. Argunov, D. Vinnitskiy, A. Gerbst, N. Ustyuzhanina, **Nikolay Nifantiev**

2:30 52. Synthetic utility of endocyclic cleavage reaction. **Shino Manabe**, Y. Ito

2:50 53. New organocatalyst for stereoselective synthesis of 2-Deoxyglycosides. **Gary Bradshaw**, A. Colgan, N. Allen, M. Boland , E. McGarrigle

3:10 372. Evaluation of the effect of L-rhamnose targeting on cellular immune responses in a liposomal MUC1 glycopeptide-based immunotherapeutic. P. Karmakar, K. Lee, K.A. Wall, **S.J. Sucheck**

3:30 55. Phenacyl ester-assisted preparation of β-Kdo glycosides: Application to synthesis of the biological repeating unit of *Burkholderia pseudomallei* exopolysaccharide. **Charles Gauthier**, M. Mazur, B. Barycza, A. Laroussarie, Y. Blériot

3:50 Intermission, Exhibits, Posters – Preservation Hall

MONDAY AFTERNOON, July 18

CARBOHYDRATE STRUCTURE, ANALYSIS & COMPUTATION
Glycoinformatics

New Orleans Marriott, Galerie 3
A. D. French, K. W. Moremen, R. J. Woods, *Organizers*; Jesus Jimenez-Barbero, *Presiding*

4:20 56. Keynote Lecture: Computing in cancer from glycobioinformatics, analytics based Glyco gene signature discovery to the simulation of glycosyltransferase mechanisms. **Kevin J. Naidoo**

4:50 57. Interoperability of glycosaminoglycans databases through semantic web. **Serge Perez**, F. Lisacek, J. Mariethoz, M. Deniau, N. Thierry-Mieg, A. Rivet, S. Ricard-Blum

5:10 58. Platforms for qualifying the features of glycans and glycomes. **Christopher B. Barnett**, K.J. Naidoo

5:30 59. Determination of substitution positions in cross-linked hyaluronic acid hydrogels by NMR spectroscopy and mass spectrometry. **Frida J. Wende**, S. Gohil, H. Mojarradi, T. Gerfaud, L. Nord, A. Karlsson, J. Boiteau, A. Helander Kenne, C. Sandström

---

GLYCOBIOCHEMISTRY & GLYCOBIOLOGY
Bacterial and Pathogen Glycans

New Orleans Marriott, Galerie 5, 6
L. L. Kiessling, J. C. Paulson, *Organizers*; C. Whitfield, *Presiding*

4:20 60. Keynote Lecture: Two 3,6-dideoxyhexoses serve as scaffolds for a combinatorial library of signaling molecules in C. elegans. **Frank Schroeder**

4:50 61. Bacterial lectins and infection – from structural glycobiology to antiadhesive strategies. **Anne Imberty**, A. Varrot, A. Titz, B. Cournoyer, S. Vidal

5:10 62. Bioengineering production of bacterial legionaminic acid for chemoenzymatic synthesis. **Dennis M. Whitfield**, C.N. Boddy, M. Hassan

5:30 63. Thermodynamics of protein and ligands conformational changes in the active site of a glycosyltransferase. J. Romero-García, **Xevi Biarnés**, D. Albesa-Jové, M.E. Guerin, A. Planas

MONDAY AFTERNOON, July 18

GLYCOBIOCHEMISTRY & GLYCOBIOLOGY
Lectins and Glycan Binding Proteins
Financially supported by Carbohydrate Research/Elsevier

New Orleans Marriott, Bissonet
L. L. Kiessling, J. C. Paulson, *Organizers*; L. C. Hsieh-Wilson, *Presiding*

4:20 64. Keynote Lecture: Precision glycopolymers as new tools to probe the multivalent binding of glycomimetics. **Laura Hartmann**

4:50 65. Dynamics of fucose and Lewis x in lectin-binding sites. **Mickael Lelimousin**, J. Topin, J. Arnaud, A. Audfray, S. Perez, A. Varrot, A. Imberty

5:10 66. Galectins, key players of homeostasis. **Cyrille Grandjean**

5:30 67. Study and application of protein O-glycosylation. **Zhongping Tan**

NEW DEVELOPMENTS IN GLYCOSCIENCE
Synthesis: Methods and Applications (2)

New Orleans Marriott, Galerie 2
A. D. French, M. A. Godshall, S. M. Murphy, J. R. Vercellotti, *Organizers*; T. Baasov, *Presiding*

4:20 68. Clustering of P^k-trisaccharide on amphiphilic cyclodextrin reaveals unprecedented affinity for the shiga-like toxin Stx2. P. Zhang, E. Paszkiewicz, J. Sadowska, K. Pavel, D.R. Bundle, **Chang-Chun Ling**

4:50 69. N-Acetylgalactosamine: A novel and flexible approach starting from tartaric acid. **Bettina Riedl**, W. Schmid

5:10 70. Semisynthesis of intact complex-type triantennary oligosaccharides from a biantennary oligosaccharide isolated from a natural source by selective chemical and enzymatic glycosylation. **Yuta Maki**, R. Okamoto, M. Izumi, T. Murase, Y. Kajihara

5:30 71. Highly selective carbohydrate based G-quadruplex ligands with anti-cancer activity. **Steven T. Street**, M. Berry, D. Chin, G. Hollingworth, J. Morales, M. Galan

MONDAY AFTERNOON, July 18

SYNTHESIS OF CARBOHYDRATES, GLYCOCONJUGATES & GLYCAN-BASED BIOMATERIALS
Synthesis of Glycomimetics

New Orleans Marriott, Acadia Room
G. Boons, M. Cloninger, A. Demchenko, *Organizers*; P. G. Wang, *Presiding*

4:20 72. Keynote Lecture: The production and inhibition of biofilm related exopolysaccharides. **Mark Nitz**

4:50 73. Using dendrimers to study the role of galectin-3 in cancer cellular processes. **Mary Cloninger**, A.K. Michel, C. Goodman, J.H. Ennist, A. Mattson

5:10 74. Glycomimetics and scaffolds from carbohydrates by highly stereoselective tandem allylic rearrangement-Huisgen cycloaddition. **Paul V. Murphy**, R. Chadda, L. Moynihan

5:30 75. Neutralization of neuropathy-inducing anti-MAG antibodies by multivalent glycoconjugates. **Helene Pfister**, P. Hänggi, F. Yang, R. Herrendorff, B. Ernst

MONDAY EVENING, July 18

ADVANCES IN THE GLYCOSCIENCES
Poster Session (1)
Financially supported by Springer

New Orleans Marriott, Preservation Hall
S. M. Murphy, *Organizer, Presiding*

5:50 - 7:00

76. Expanding semisynthesis of glycoprotein by sugar-scaffolded hydrophilic protecting group. **C. Chandrashekar**, R. Okamoto, M. Izumi, Y. Kajihara

77. Chemi-enzymic synthesis of isotopically labeled nucleosides from a common labeled precursor. **W. Zhang**, T. Turney, I. Surjancev, A.S. Serianni

78. A chemical synthesis of a multiply ^{13}C-labeled hexasaccharide: A high-mannose *N*-glycan fragment. **W. Zhang**, Q. Pan, A.S. Serianni

79. Synthesis of assorted 6-*O*-acyl-α-glucosyl cholesterols from *Helicobacter* spp. using a highly stereoselective glucosylation. **D. Smith**, S.J. Williams

MONDAY EVENING, July 18

80. Regioselective synthesis of high-mannose oligosaccharides with minimally protected glycosyl acceptors. **Q. Pan**, Q. Wang, B. Meng, J. Wang, A.S. Serianni

81. *O*-Acetyl groups in saccharides: NMR *J*-couplings and statistical models for acetate ester conformational analysis. T. Turney, **Q. Pan**, I. Carmichael, A.S. Serianni

82. Synthesis and immunomodulatory activity of caryophyllose-containing lipooligosaccharides. **C. Chu**, T.L. Lowary

83. Synthesis of defined heparan sulfate fragments and sulfated pseudo-oligosaccharide GAG mimetics as inhibitors of Enterovirus 71. **D. Earley**, B. Bailly, R. Thomson, C. Chang, M. von Itzstein

84. Synthesis of photoaffinity probes via triple-click reaction and their applications in protein labeling. **T. Lin**, C. Lin

85. Chemo-enzymatic synthesis and biological evaluation of bacterial sialidase inhibitors. **T.J. Slack**, Y. Li, D. Shi, M.M. Muthana, Y. Liu, X. Chen

86. Chemoenzymatic synthesis of glycolipids with facile purification. **A. Santra**, Y. Li, H. Yu, X. Chen

87. Facile enzymatic synthesis of *N*-glycans with tandem epitopes based on isolated natural core glycan. Z. Wu, **Y. Liu**, C. Ma, Y. Guo, W. Liuqing, A. Calderon, H. Zhu, X. Chen, W. Guan, **P.G. Wang**

88. An efficient orthogonal strategy for the synthesis of high mannose *N*-glycans. **Y. Liu**, J. Zhou, X. Cao, P.G. Wang

89. Facile chemoenzymatic synthesis of a library of human milk oligosaccharides. **Y. Guo**, Z. Xiao

90. Efficient synthesis of specialty carbohydrates through industrial biotechnology. **J. Beauprez**, W. Soetaert

91. 6-Methylpyrid-2-yl glycosides and their role in oligosaccharide synthesis. **J. Causey**, **A. Trujillo**, **S. Horstmeier**, **V. Smith**, **S. Hasty**

92. Chemoenzymatic synthesis of asymmetric biantennary *N*-glycan library: Influence of counter-anion on stereoselectivity of key chemical glycosylations. **J. Zhou**, X. Cao, Y. Liu, H. Zhu, A. Calderon, L. Li, P.G. Wang

93. Synthesis and evaluation of thiophenyl sialyl donor bearing picoloyl at C-4: Triflic acid effect. **S. Escopy**, S. Geringer, F. Najafi, C. De Meo

MONDAY EVENING, July 18

94. Anthracycline glycodiversification: semi-synthesis of novel doxorubicin glycoside. **M.B. Martins-Teixeira**, A.P. Corrado, I. Carvalho

95. Stereoselective synthesis of new lipophilic carba analogues of minimal epitope Manα(1,2)Man as potential DC-SIGN ligands. S. Fortunato, A. Marsico, P. Crotti, V. Di Bussolo, **V. Bordoni**

97. Synthesis and immunomodulatory functions of mycobacterium peptidoglycan fragments. Q. Wang, Y. Matsuo, K. Fukase, **Y. Fujimoto**

98. Synthesis of a MUC1-glycopeptide antitumor vaccine linked to a mannose ligand for macrophages. **M. Glaffig**, N. Stergiou, E. Schmitt, H. Kunz

99. Semi-synthesis of α2,6-sialyltransferase bearing high mannose-type oligosaccharide. **A. Shimada**, R. Okamoto, M. Izumi, Y. Kajihara

100. Glycosylation reaction of sulfated donor for the synthesis of homogenous sulfated oligosaccharides. **A. Manbo**, R. Okamoto, M. Izumi, Y. Kajihara

101. Establishing a route for the synthesis of ipomoeassin F analogs containing a modified aglycone. **E. Barber**, W. Shi, G. Zong

102. Antibacterial deoxy glycosides: Synthesis and toxicity. A.I. Martins, C. Dias, J.P. Pais, A.P. Rauter, **J.A. Justino**

103. A second-generation synthesis of the immunogenic mycobacterial glucuronosyl diacylglyceride Gl-A. **S. Burugupalli**, S. Shah, M.B. Richardson, S.J. Williams

104. The synthesis of α-mannosyl-1,3-noeuromycin: A potent inhibitor of *endo*-α-1,2-mannanase. **M. Petricevic**, L. Sobala, G. Davies, S.J. Williams

105. Transforming flask reaction into *Escherichia coli* for the synthesis of polyhydroxylated molecules based on DHAP-dependent aldolases. **Z. Li**, M. Wei, L. Cai, P.G. Wang

106. Design, synthesis, and activity of nanocellulosic protease sensors. **V. Edwards**, K.R. Fontenot, N.T. Prevost, A.D. French, B.D. Condon, F. Liebner, D. Haldane

107. Total synthesis of TMG-chitotriomycin based on electrochemical automated solution-phase synthesis of the precursor tetrasaccharide. **Y. Isoda**, N. Sasaki, K. Kitamura, T. Nokami, T. Itoh

MONDAY EVENING, July 18

108. Regioselective opening of 3,5-*O*-di-*tert*-butylsilylene-D-galactofuranosides: Synthesis of (1→5)-β-D-galactofuran derivatives. **C. Gallo-Rodriguez**, M.J. Tilve

109. Synthesis of Gal*p*(α1→3) [Gal*p*(β1→6)]Gal*p*, a model for studying the interplay between the anti-α-Gal*p* antibody and the trans-sialylation in the infectivity of *Trypanosoma cruzi*. **C. Marino**, M. Giorgi, R.M. Lederkremer

110. Glycosylation reactions with α- and β-D-gluco and D-galactopyranosyl acceptors: Experimental and theoretical study. **C. Marino**, C. Stortz, E. Del Vigo

111. Efficient synthesis of *N*-glycans containing core fucose or bisecting GlcNAc. **Y. Manabe**, M. Nagasaki, H. Syomura, N. Minamoto, K. Fukase

112. Synthetic efforts toward the synthesis of the hexasaccharide O-antigen of *Vibrio cholerae* O139. **S.E. Soliman**, P. Kováč

113. Synthesis of oligosaccharide mimics as acceptor substrates of the novel galactosyltransferase GalT-1. **M.B. Calvert**, A. Titz

114. Semi-synthesis of bioactive glycosylphosphatidyl-inositols anchored protein from *P. falciparum*. **A. Malik**, B. Lee, P.H. Seeberger, D. Varon Silva

115. Towards the synthesis of O-acetylated fragments of Salmonella Typhi Vi capsular polysaccharide and their zwitterionic analogues. **L. Lay**, M. Sella, M. Fusari, D. Cancogni, F. Maggioni

116. The first synthesis of the pentasaccharide fragment of the *O*-specific antigen of *Vibrio cholerae* O139. **X. Lu**, P. Kováč

117. Stereoselective synthesis of spiro bis-*C,C*-α-arylglycosides by tandem Heck type *C*-glycosylation and Friedel–Crafts cyclization. **H. Lin**

118. A facile and efficient synthesis of the α-Gal trisaccharide using an Fmoc-based strategy. **K.R. Bartholomew**, G.J. McGarvey

119. Synthesis of ABC-transporter dependent pathway oligosaccharides from *Klebsiella pneumonia*. **R. Sweeney**, T.L. Lowary

120. Synthesis of lipopolysaccharide O-chain biosynthetic fragments from *Escherichia coli* O9A. **L. Wang**, T.L. Lowary

121. Studies towards the synthesis of rare sugar building blocks containing campylobacter glycans. **V. Thota**, **T. Lowary**

MONDAY EVENING, July 18

122. Metabolic labeling of lipopolysaccharide in Gram-negative bacteria: Synthesis, purification, and characterization of Kdo-N_3. **I. Nilsson**, N.N. Kreamer, D. Tang, T. Uehara, G. Lapointe, D.A. Six

123. Catalytic stereoselective synthesis of glycosides. **C. Palo-Nieto**, A. Sau, R. Williams, M. Galan

124. Synthesis of analogues of decaprenolphosphoarabinose (DPA) as potential anti-TB agents. **K. Suthagar**, A.J. Fairbanks

125. Design and synthesis of a high-throughput fluorogenic substrate for Endo-*β*-*N*-acetylglucosaminidase. **N. Ishii**, C. Sunaga, Y. Yoshimura, J. Kumada, Y. Matsuzaki, I. Matsuo

126. Carbohydrates synthesis based on lactol dynamic kinetic transformations. **H. Wang**

127. Synthesis of novel carbohydrate compounds for potential CNS-drugs. **E. Hanawa-Romero**, C.E. Marzabadi, I.J. Talisman

128. An inner membrane complex is essential for the biosynthesis of the Pel polysaccharide in *Pseudomonas aeruginosa*. **G. Whitfield**, J.D. Rich, L.S. Marmont, M.R. Parsek, J.J. Harrison, L. Howell

129. Structure-affinity relationships in glycan recognition: Bacteriophage tailspike proteins to explore the role of solvent in carbohydrate binding. U. Gohlke, N.K. Broeker, S. Kunstmann, M. Santer, **S. Barbirz**

130. Immunomodulatory activity of chemically synthesized lipopolysaccharide partial structures. **A. Shimoyama**, Y. Fujimoto, K. Fukase

131. Soluble dietary fiber: Relationship between chemical structure and fermentation profile by the human intestinal microbiota *in vitro*. T. Jungles, **L.M. Cordeiro**, M. Iacomini, B. Hamaker

132. The relationship between glucosamine 3-*O* sulfotransferase efficiency and substrate structure on bioengineering for anticoagulant heparan sulfate. **T. Ono**, S. Yamazaki, Y. Tokura, Y. Mihara, N. Sugiura, K. Mori

133. Structure-reactivity relationships of conformationally armed glucosyl donors. **T. Angles d'Ortoli**, C. Hamark, G. Widmalm

134. A common glycan structure on immunoglobulin G for enhancement of effector functions. **C. Lin**, M. Tsai, S. Li, T. Tsai, K. Chu, Y. Liu, M. Lai, C. Wu, Y. Tseng, S.S. Shivatare, C. Wang, P. Chao, S. Wang, H. Shih, Y. Zeng, T. You, J. Liao, Y. Tu, Y. Lin, H. Chuang, C. Chen, C. Tsai, C. Huang, N. Lin, C. Ma, C.Y. Wu, C. Wong

MONDAY EVENING, July 18

---

135. Structurally close diastereomeric ligands induce different solvation and conformational entropy alterations upon binding to galectin-3. M.L. Verteramo, F. Manzoni, O. Stenström, M.M. Ignjatovic, M. Olsson, H. Leffler, U. Ryde, D. Logan, M. Akke, U.J. Nilsson

136. Synthesis of novel thiodigalactoside derivatives with improved affinity for galectin-3. **K. Peterson**, P. Verma, F. Zetterberg, U.J. Nilsson

137. Synthesis and evaluation of octyl glycosides for membrane protein analysis. T. Saito, S. Hasemi, M. Ishikawa, Y. Yoshimura, **I. Matsuo**

138. Rapid and sensitive MALDI MS analysis of oligosaccharide by using 2-hydrazinopyrimidine as a derivative reagent and co-matrix. **K. Jiang**, P.G. Wang

139. Quantitative analysis of the expression of hyaluronan/chondroitin sulfate-degrading enzymes in the mouse brain during development. **K. Miwa**, S. Mizumoto, S. Yamada

140. 2-hydrazinoquinoline as a novel analytical weapon for N-glycan analysis and characterization. **A. Aloor**, K. Jiang, P.G. Wang

141. Deuterated carbohydrate probes as 'label-free' substrates for probing nutrient uptake in mycobacteria by nuclear reaction analysis. **E. Fullam**

142. Structural and functional analysis of the solute-binding protein UspC from *Mycobacterium tuberculosis* that is specific for amino sugars. **E. Fullam**

143. Functional analysis of an endomannosidase in the endoplasmic reticulum: triaging Glc1Man9GlcNAc2-glycoproteins. **M. Hirano**, C. Watanabe, Y. Ito, S.J. Williams, K. Totani

144. Structural analysis and anti-obesity effect of a pectic polysaccharide isolated from mulberry fruit. **J. Choi**, A. Synytsya, P. Capek, R. Bleha, R. Pohl, Y. Park

145. Anti-cancer effects of AEG-1, a glycosylated phenolic compound, in human lung cancer cells. **J. Choi**, T. Nguyen, J. Sohng, Y. Park

146. Application of zwitterionic-type hydrophilic interaction liquid chromatography coupled with electrospray ionisation mass spectrometry (ZIC®-HILIC-ESI-MS) for separation and analysis of lipopolysaccharide-derived oligosaccharides. A. Man-Kupisinska, E. Bobko, **A. Maciejewska**, T. Gozdziewicz, W. Jachymek, C. Lugowski, J. Lukasiewicz

147. Carbohydrate-derived oligo(amide-triazoles) as anion binding receptors. NMR and MS analysis. D. Fidalgo, M.E. Monge, **A. Kolender**, O. Varela

MONDAY EVENING, July 18

148. Analysis of amino acid residues relating to specificity of endodextranase from *Streptococcus mutans*. P. Klahan, M. Okuyama, H. Mori, A. Kimura

149. Structural and thermodynamic analysis of chitin oligosaccharide recognition by a family 50 carbohydrate-binding module. **T. Ohnuma**, T. Taira, **N. Umemoto**, Y. Kitaoku, T. Numata, T. Fukamizo

150. Green synthesis of silver nanoparticles using natural dyes of cochineal. **B. Kumar**, L.H. Cumbal

151. Isolation and partial purification of a lectin from the leaves of *Chrysophyllum Cainito*. **D. M**, E. K K

152. Interaction of pancreatic α-amylase with its identified glycoligand SGLT1. **K. Date**, A. Satoh, H. Ogawa

153. Aminosequence for H-type II specific lectin from *Galactia lindenii s*eeds and production of recombinant lectin in a bacterial system. **C. Nury**, M. Quintero, E. Reyes, E.A. Reyes, N.A. Vega

154. Purification of a new lectin present in seeds from *Lepechinia bullata and Salvia bogotensis* genus. **A. Wilches**, F. José Luis, E.A. Reyes, A. Imberty, A. Varrot, N.A. Vega

155. Isolation and characterization of novel lectin from *Galactia lindenii* seeds. **C. Zulma**, M. Quintero, E.A. Reyes, N.A. Vega

158. 2-Deoxy glycosides as innovative antimicrobial and neuroprotective molecules. **C. Dias**, A.P. Rauter

159. Design, synthesis and selective T-cell activations of 5-Thio-α-GalCer analogues. **Yuguo Du**

160. Convergent synthesis of O-glycan core structures and enzymatic extension of O-glycan library. **G. Edmunds**, M. Wei, Q. Zhang, L. Li, P.G. Wang

161. Total synthesis of the *Morganella morganii* zwitterionic polysaccharide repeating unit. **D.J. Keith**, S.D. Townsend

162. Synthesis of oligosaccharide fragments of rhamnogalacturonan-II. **E. Mancuso**, M. Clausen

163. Understanding the functional mechanism of ipomoeassin natural glycoresins. **W. Shi**, G. Zong, Z. Hu, H. Aljewari, J. Zhou, Y. Du

164. Synthesis of pseudaminic acid for the study of microbial glycobiology in the human gut. T. Sminia, H. Zuilhof, T. Wennekes

165. Synthesis of bacterial mimetics of HIV-1 GP120 glycans. **N. Trattnig**, R. Pantophlet, P. Mayrhofer, R. Kunert, P. Kosma

166. Transglycosylation reaction of bacterial *N*-acetyl hexosaminidases towards human milk oligosaccharide core structures. **T. Visnapuu**, T. Katayama, A. Lie, L.H. Pedersen, P. Stougaard, B. Svensson

167. Ni-Catalyzed Stereoselective Glycosylation for the formation of 1,2-cis-2-Amino Glycosides and Applications. **F. Yu**

168. A recombinant α-fucosidase is capable of core fucosylation of GlcNAc-peptides. **C. Li**, L. Wang

169. Effects of cleavage of high-mannose-type glycans of *Maackia amurensis* leukoagglutinin on sialic acid-binding activity. **H. Park**, J. Kim, H. Oh, J. Choi, Y. Lee, H. Kim

170. Cleavage of the terminal N-acetylglucosamine of ovalbumin N-glycans significantly reduces IgE production and Th2 cytokine secretion. **J. Kim**, **H. Park**, H. Oh, J. Choi, Y. Lee, H. Park, H. Choi, H. Kim

171. Glycan structure and serum half-life of recombinant CTLA4Ig, an immunosuppressive agent, expressed in rice cells with coexpression of human β1,4-galactosyltransferase and human CTLA4Ig. **H. Park**, **J. Kim**, H. Oh, J. Choi, Y. Lee, H. Kim

172. Semi-synthetic glycosaminoglycan analogues as potential therapeutics for Alzheimer's diseases. L. Hadfield, J. Turnbull, E. Yates, **M. Skidmore**

173. Glucopyranos-6'-yl purines and pyrimidines as novel types of isonucleosides possessing therapeutic potential. **D. Batista**, S. Schwarz, A. Loesche, R. Csuk, R. Jorda, P. Costa, N. Xavier

174. New generation of UDP-peptide conjugates as bisubstrate cell-penetrating OGT inhibitors. **V. Borodkin**, K. Rafie, A. Gorelik, R. Trapannone, D. van Aalten

TUESDAY MORNING, July 19

PLENARY SESSION
Plenary Lecture (4)

New Orleans Marriott, Acadia Room
S. M. Murphy, *Organizer*; S. J. Eichhorn, *Presiding*

8:30 175. Glycans and glyconanotechnology. **Soledad Penadés**

PLENARY SESSION
Plenary Lecture (5)

New Orleans Marriott, Acadia Room
S. M. Murphy, *Organizer*; Y. Ito, *Presiding*

9:15 176. Structural views of glycan-dependent determination of glycoprotein fates in cells. **Koichi Kato**

10:00 Intermission, Exhibits, Posters – Preservation Hall

CARBOHYDRATE STRUCTURE, ANALYSIS & COMPUTATION
Spectroscopy and Spectrometry

New Orleans Marriott, Galerie 3
A. D. French, R. J. Woods, *Organizers* ; K. W. Moremen, *Organizer, Presiding*

10:30 177. Keynote Lecture: A mass spectrometry view into diverse aspects of glycobiology. **Albert J. Heck**

11:00 178. NMR and MS structural analysis of glycocongugates on demand at the Complex Carbohydrate Research Center. **Parastoo Azadi**

11:20 179. Structural analysis of *Klebsiella pneumoniae* O-antigens facilitated by high-resolution magic-angle spinning (HR-MAS) ^{1}H, ^{13}C NMR spectroscopy. K. Stojkovic, M. Kaszowska, T. Niedziela, C. Lugowski, **Jolanta Lukasiewicz**

11:40 180. Raman optical activity of saccharides. **Jakub Kaminsky**, V. Profant, V. Kopecký, J. Šebestík

12:00 181. Spectroscopy-enabled comprehensive *de novo* sequencing of underivatised carbohydrates. **Isabelle Compagnon**

TUESDAY MORNING, July 19

GLYCOBIOCHEMISTRY & GLYCOBIOLOGY
Bacterial Glycans

New Orleans Marriott, Bissonet
L. L. Kiessling, J. C. Paulson, *Organizers*
A. Planas, *Presiding*

10:30 182. Keynote Lecture: The structure and functions of bacterial β-Kdo glycosyltransferases. **Chris Whitfield**

11:00 183. The acceptor binding epitope of heptosyltransferase I revealed. WaaC caught in the act! **Markus Blaukopf**, P. Kosma, L. Worrall, N. Strynadka, S. Withers

11:20 184. Structure and mechanism of action of peptidoglycan *O*-acetyltransferase A (OatA) from *Streptococcus pneumoniae*. **David Sychantha**, D. Little, L. Howell, A.J. Clarke

11:40 185. The evolution of carbohydrate-active enzyme specificity in genomes. N. McGregor, V. Yin, C. Tung, F. van Petegem, **Harry Brumer**

12:00. Bruce A. Stone Award, to be presented by Vincent Bulone

GLYCOBIOCHEMISTRY & GLYCOBIOLOGY
Proteoglycans and Cell Surface Glycans

New Orleans Marriott, Galerie 5, 6
L. L. Kiessling, J. C. Paulson, *Organizers*; Y. Li, *Presiding*

10:30 187. Keynote Lecture: Hyaluronan and hyaluronan fragments. **Mary K. Cowman**

11:00 188. Cell surface glycan engineering with well-defined heparan sulfate derivatives through selective exo-enzymatic labeling. **Chantelle J. Capicciotti**, C. Zong, G. Boons

11:20 189. Using re-engineered lectins to construct proto-tissues. **Daniel J. Williamson**, T.R. Branson, S. Villringer, J. Madel, W. Römer, B. Turnbull

11:40 190. Glycocalyx based strategies to modulate embryonic stem cell fates. **Mia Huang**, C.J. Fisher, A. Michalak, M. Christy, K. Godula

12:00 191. Contribution of PECAM-1 glycans to endothelial cell barrier function. **Panida Lertkiatmongkol**, D. Zhou, D.K. Newman, J. Zhu, P.J. Newman

TUESDAY MORNING, July 19

NEW DEVELOPMENTS IN GLYCOSCIENCE
Biosynthesis and Synthesis

New Orleans Marriott, Galerie 2
A. D. French, M. A. Godshall, S. M. Murphy, J. R. Vercellotti, *Organizers*; X. Huang, *Presiding*

10:30 192. Sph3, a glycoside hydrolase required for galactosaminogalactan biosynthesis by *Aspergillus fumigatus*. **Natalie C. Bamford**, B.D. Snarr, F.N. Gravelat, D. Little, P. Baker, M.J. Lee, H. Robinson, D.C. Sheppard, L. Howell

11:00 193. Influence of acyl group on Lewis acid promoted anomerisation reactions. **Paul V. Murphy**, L. Kerins, M.P. Farrell

11:20 194. Synthesis and insights onto the key structural features of nucleoside-based inhibitors of butyrylcholinesterase. **Vasco Cachatra**, A.P. Rauter

11:40 195. Substrate specificity of FUT8 and chemoenzymatic synthesis of core-fucosylated asymmetric *N*-glycans. **Angie D. Calderon Molina**, P.G. Wang, L. Li

12:00 196. Chemoenzymatic approach to asymmetrically bisected *N*-glycan library. **Xuefeng Cao**, Jun Zhou, Y. Liu, A. Calderon, L. Li, P.G. Wang

SYNTHESIS OF CARBOHYDRATES, GLYCOCONJUGATES & GLYCAN-BASED BIOMATERIALS
Synthesis of Glycoconjugates

New Orleans Marriott, Acadia Room
G. Boons, A. Demchenko, *Organizers*; M. Cloninger, *Organizer, Presiding*

10:30 197. Keynote Lecture: Space, time and glycan processing: Synthetic and analytical approach. **Osamu Kanie**

11:00 198. Synthesis of *Neisseria meningitidis* Y tetramer glycoconjugate and its immunogenicity studies. M. Chhikara, **Kishore R. Harale**, R. Rana, J. Dalal, S. Hanif, N. Kumar

11:20 199. Semi-synthesis and properties of homogeneous glycosylphosphatidylinositol anchored proteins. R. Roller, M. Carillo, M. Garg, D. Michel, P.H. Seeberger, **Daniel Varon Silva**

11:40 200. Synthesis of elongated mucin core glycopeptides to decipher lectin and antibody binding recognition events . C. Pett, M. Schorlemer, **Ulrika Westerlind**

12:00 201. Chemical synthesis of glycoproteins for understanding of glycoprotein quality control system. **Yasuhiro Kajihara**, T. Kiuchi, M. Izumi, Y. Takeda, A. Seko, M. Sakono, Y. Ito

12:20 – 1:30 LUNCH

TUESDAY AFTERNOON, July 19

CONVERSION & UTILIZATION OF LIGNOCELLULOSIC BIOMASS

New Orleans Marriott, Galerie 5, 6
S. M. Murphy, *Organizer*; N. Labbé, *Organizer, Presiding*

1:30 202. Keynote Lecture: Synthetic biology of plant cell wall polysaccharides. **Markus Pauly**

2:00 203. Characterization of plant cell wall remodeling and degrading enzymes using synthetic oligosaccharides. **Colin Ruprecht**, D. Schmidt, P. Dallabernardina, F. Pfrengle

2:20 204. Identification and characterization of a novel fungal ascomycete laccase. **Majid Haddad Momeni**, P. Bollella, R. Antiochia, L. Gorton, M. Abou Hachem

2:40 206. Pyrolysis bio-oil as precursor of polymeric bio-based resins. **Maria L. Auad**, B. Sibaja, M. Barde

3:00 205. Can carbohydrates reduce our reliance on fossil fuels? **Indra N. Pulidindi**, A. Gedanken

3:20 Intermission, Exhibits, Posters – Preservation Hall

DEVELOPMENT & COMMERCIALIZATION OF CUTTING-EDGE ANALYTICAL TECHNOLOGIES FOR CHARACTERIZATION OF CARBOHYDRATES
Vendor Updates

New Orleans Marriott, Galerie 3
S. M. Murphy, *Organizer*; M. A. Godshall, *Organizer, Presiding*

1:30 207. Keynote Lecture: The dynamic state of glycan mass spectrometry. **Catherine E. Costello**

2:00 208. NMR analysis of carbohydrates: Challenges and successes. **Clemens Anklin**

2:20 209. RapiFluor-MS labeled glycan analysis for biotherapeutic proteins. **William Alley**

2:40 210. The characterization of fractionated polysaccharides with multi-angle light scattering (MALS). **Jeffrey Ahlgren**

3:00 211. Comprehensive N-glycosylation analysis of biopharmaceuticals by capillary electrophoresis and mass spectrometry. **Andras Guttman**

3:20 Intermission, Exhibits, Posters – Preservation Hall

TUESDAY AFTERNOON, July 19

GLYCOBIOCHEMISTRY & GLYCOBIOLOGY
Designing Probes of Glycan Function

New Orleans Marriott, Bissonet
L. L. Kiessling, J. C. Paulson, *Organizers*; M. I. Gibson, *Presiding*

1:30 212. Keynote Lecture: The conformational variability of FimH – which conformation represents the therapeutic target? **Beat Ernst**

2:00 213. Molecular design and mechanistic study of aminosugar-derived bacterial and fungal cell disruptors. **M. Fridman**

2:20 214. *C*-Glucosyl dihydrochalcones as potent and selective SGLT2 inhibitors: Synthesis and bioactivity. **A.P. Rauter**, A.R. Jesus, T.M. Dore

2:40 215. Carbohydrate-based inducers of cellular stress for targeting cancer cells. **F. Ndombera**

3:00 216. Enzymatic thioligation: En route to the preparation of glycosyl amino acids. L. Guillotin, P. Lafite, **R. Daniellou**

3:20 Intermission, Exhibits, Posters – Preservation Hall

NEW DEVELOPMENTS IN GLYCOSCIENCE
Carbohydrates and Enzymes

New Orleans Marriott, Galerie 2
A. D. French, M. A. Godshall, S. M. Murphy, J. R. Vercellotti, *Organizer;* C. E. Marzabadi, *Presiding*

1:30 217. Evolution of receptor specificity in seasonal and pandemic influenza. **Andrew Thompson**, W. Peng, R. de Vries, P. Lee, R. McBride, I.A. Wilson, J.C. Paulson

2:00 218. Degradation pathway of plant complex-type *N*-glycans: Identification and characterization of α1,3-fucosidase acting on plant complex type *N*-glycans. S. Kato, M.

Hayashi, M. Kitagawa, M. Maeda, Y. Kimura, K. Igarashi, M. Samejima, M. Kasahara, **Takeshi Ishimizu**

2:20 219. Genetically-encoded selection of glycopeptide agonists for carbohydrate binding proteins. **Ratmir Derda**

2:40 220. Determination of sugar metabolism profiles for non-traditional brewing yeasts of the genus *Brettanomyces*. **Kent Johnson**, W. Deutschman

3:00 221. A convergent approach to HMO mimics. **Dorothy L. Ackerman**, S.D. Townsend

3:20 Intermission, Exhibits, Posters – Preservation Hall

TUESDAY AFTERNOON, July 19

SYNTHESIS OF CARBOHYDRATES, GLYCOCONJUGATES & GLYCAN-BASED BIOMATERIALS

Synthesis and Recognition

New Orleans Marriott, Acadia Room
G. Boons, M. Cloninger, A. Demchenko, *Organizers*; P. R. Andreana, *Presiding*

1:30 222. Keynote Lecture: Synthetic C-lectins: A new route for targeting cell-surface glycans. **Adam B. Braunschweig**, Y. Zheng, Y. Zhou

2:00 223. Targeting siglec receptors to avoid the autoimmune destruction of β-cells in type 1 diabetes. **Rachel Hevey**, B. Ernst

2:20 96. Synthesis of cell-surface heparan sulfate saccharides for probing specific interaction with proteins. **Shang-Cheng Hung**

2:40 225. Application of the inverse-electron-demand Diels-Alder reaction for the on-chip screening of glycosylation promoters. **Valentin Wittmann**, C. Risinger

3:00 226. Synthesis of plant oligosaccharides for protein-binding studies. **Irene Boos**, M. Clausen

3:20 Intermission, Exhibits, Posters – Preservation Hall

TUESDAY AFTERNOON, July 19

DEVELOPMENT & COMMERCIALIZATION OF CUTTING-EDGE ANALYTICAL TECHNOLOGIES FOR CHARACTERIZATION OF CARBOHYDRATES
Characterization Tools for Development

New Orleans Marriott, Galerie 3
M. A. Godshall, S. M. Murphy, *Organizers* H.N. Cheng, *Presiding*

3:50 227. **Keynote Lecture:** Expanding the imaging toolbox. **Jennifer A. Prescher**

4:20 228. Multi-detector hollow-fiber flow field-flow fractionation (HF5) of polysaccharides. **Andre M. Striegel**, L. Pitkanen

4:40 229. Absolute quantification of N-glycans by isotopic dilution. **Juan Echevarria**, B. Echeverria, N. Ruiz, A. Hernandez, J. Calvo, N.C. Reichardt

5:00 230. Novel enzyme substrates: Transforming the way we measure *endo*-acting glycosyl hydrolases. **David Mangan**, C. Cornaggia, B. McCleary

GLYCOBIOCHEMISTRY & GLYCOBIOLOGY
Glycan Engineering
Financially supported by Carbohydrate Research/Elsevier

New Orleans Marriott, Bissonet
L. L. Kiessling, J. C. Paulson, *Organizers* ; L. Hartmann, *Presiding*

3:50 231. **Keynote Lecture:** Liposome-assisted metabolic glycan labeling *in vivo*. **Xing Chen**

4:20 232. Chemical reporters for the investigation of O-GlcNAc glycosylation and carbohydrate metabolism. **Matthew Pratt**

4:40 233. O-GlcNAc modification inhibits apoptosis through direct glycosylation of caspases. **Kelly N. Chuh**, A. Batt, N. Marotta, C. Brennan, M. Pratt

5:00 234. Precision editing of the cellular glycocalyx with nanoscale glycomaterials. **Kamil Godula**

TUESDAY AFTERNOON, July 19

GLYCOTECHNOLOGY IN MEDICINE, NUTRITION & MATERIAL SCIENCE
Glycotechnology in Nutrition

New Orleans Marriott, Galerie 5, 6
M. A. Godshall, S. M. Murphy, M. Roman, *Organizers*
S. Austin, *Presiding*

3:50 235. Keynote Lecture: Potato digestibility: Composition, structure and function of cell walls. **J.K. Frost**, B. Flanagan, M. Gidley, D. Rosendale, D. Brummell, J. Monro

4:20 236. Structure and enzyme digestibility of resistant starch. Y. Ai , H. Jiang, **J. Jane**

4:40 237. Comparison of anti-pathogenic activities of the human and bovine milk N-glycome: Fucosylation is a key factor. W. Wang, Y. Du, Z. Cai, A. Kulinich, S. Liu, L. Liu, **J. Voglmeir**

5:00 238. Sulfated polysaccharides from red microalgae as novel emulsifiers in food emulsions. **I. Dvir**, N. Yehuda , S. Arad (Malis)

NEW DEVELOPMENTS IN GLYCOSCIENCE
Lignocellulose Utilization

New Orleans Marriott, Galerie 2
A. D. French, M. A. Godshall, S. M. Murphy, *Organizers*; M. S. Peresin, *Presiding*

3:50 239. Structural enzymology of beta-mannanases and enzymatic synthesis of alkyl glycoside surfactants based on hemicellulose. **Henrik Stalbrand**

4:20 240. Solar-energy driven bioethanol production from carbohydrates for transportation applications. **Betina Tabah**, I.N. Pulidindi, V. Chitturi, L.M. Arava, A. Gedanken

4:40 241. A metagenomic approach to discover a novel beta-glucosidase from bovine rumens. **Prachumporn Kongsaeree**, E. Suwan

5:00 242. Cellulose-bodipy nanohybrids for singlet oxygen production. **Prashant Chauhan**, N. Yan

TUESDAY AFTERNOON, July 19

SYNTHESIS OF CARBOHYDRATES, GLYCOCONJUGATES & GLYCAN-BASED BIOMATERIALS
Chemical Sialylation

New Orleans Marriott, Acadia Room
G. Boons, M. Cloninger, A. Demchenko, *Organizers*; A. J. Fairbanks, *Presiding*

3:50 243. Keynote Lecture: The effect of O-substituents on the outcome of chemical sialylation: H-Bond-mediated aglycone delivery and beyond. **Cristina De Meo**

4:20 244. Progress toward the synthesis of legionaminic and pseudaminic acid glycosides. B. Dhakal, O. Popik, **David Crich**

4:40 245. Synthetic study on unique sialic acid containing glycan architectures of neuritogenic gangliosides found in echinoderms. **Hiromune Ando**, H. Tamai, T. Suzuki, A. Imamura, H. Ishida, M. Kiso

5:00 246. Synthesis of Kdo oligomers using 3-iodo-Kdo fluoride donors. B. Pokorny, P. Gritsch, **Paul Kosma**

TUESDAY EVENING, July 19

Advances in the Glycosciences
Poster Session (2)
Financially supported by Springer

New Orleans Marriott, Preservation Hall
S. M. Murphy, *Organizer, Presiding*

5:20 - 6:40

247. Bovine heparin: Analysis, reintroduction, and potential dangers. **K. St.Ange**, X. Liu, L. Lin, Y. Chen, F. Zhang, R.J. Linhardt

248. Development of an instant glycan labeling dye for high throughput analysis by mass spectrometry. M. Kimzey, Z. Szabo, V. Sharma, A. Gyenes, S. Tep, **A. Jones**, A. Taylor, J. Hyche, T. Haxo, S. Vlasenko

249. Protective effects of a glycosylated phenolic compound on acute pancreatitis *in vitro*. J. Seo, **S. Cho**, Y. Park

TUESDAY EVENING, July 19

---

250. Immunostimulating and anti-cancer activities of a water-soluble polysaccharide from a defatted edible nut cake. **S. Cho**, J. Choi, Y. Park

251. Saccharide primer detects glycolipid biosynthetic trend in renal cell carcinoma. **C. Kikuchi**, M.Z. Kasuya, A. Tsuchida, A. Matsuda, K. Hatanaka

252. Investigation of the amino acid residues in hyaluronidases-1 and -4 responsible for their substrate recognition. **Y. Abe**, K. Yamagiwa, S. Mizumoto, S. Yamada

253. Development of β-stereoselective mannosylations using glycosyl donors bearing 2,6-lactones. **Y. Hashimoto**, S. Tanikawa, R. Saito, K. Sasaki

254. A new role of pancreatic lipase glycosylation found by comparison of recombinant lipase expressed in *E. coli* with native lipase. **N. Kawaguchi**, C. Tomita, R. Naradate, K. Aikawa, H. Ogawa

255. Characterization of a β-1,3 specific β-L-arabinofuranosidase from *Bifidobacterium longum* subsp. *longum*. **K. Fujita**

256. Studies on the mechanism of degradation of chondroitin sulfate/hyaluronan in central nervous system. **S. Yamada**, S. Nagai, S. Mizumoto

257. Automated glycan assembly of xyloglucan oligosaccharides. **P. Dallabernardina**, F. Schuhmacher, P.H. Seeberger, F. Pfrengle

258. Chemical sulfation of glycosaminoglycan in aqueous solvent. **K. Takeda**, N. Asami, T. Minamisawa, K. Suzuki, H. Fujita

259. Preparation of electrospun nanofibers of alginate/ionic liquid blended solution. S. Park, **T. Oh**, G. Kim, Y. Kim

260. Applied tensile force determines lectin specificity. **C. Fessele**, T. Gutsmann, T.K. Lindhorst

261. Louisiana alligator processing by-products are a source of potentially bioactive glycosaminoglycans. **J. Estrada Andino**, J.N. Losso, J. King, M. Shirley, R.A. Laine, G.L. Waldrop, Y. Saini

262. Dual thio-digalactoside-binding modes of human galectins as the structural basis for the design of potent and selective inhibitors. **H. Lin**, P. Dashnyam, C.H. Lin

263. Reaction mechanism of glycosidases by studying 1,5-anhydro-d-fructose (AF) production. **J. Maneesan**, H. Matsuura, T. Tagami, H. Mori, A. Kimura

264. The USDA starch research method: A rapid method for total, soluble, and insoluble starch quantification. M. Cole, **G. Eggleston**, A. Triplett

266. New insights for enzyme engineering of family GH20 hexosaminidases: Lacto-N-biosidase as model enzyme. C. Val, X. Biarnés, M. Faijes, **A. Planas**

267. Subtle structural changes effect the metabolic fate of chemical reporters of glycosylation. **A. Batt**, K.N. Chuh, B. Zaro, M. Pratt

268. HEXPIN: Hetero-exopolysaccharide – milk protein interaction. **J. Birch**, H.K. Kári Hardarson, S. Khan, E.P. Stender, B.B. Kragelund, M. Van Calsteren, C. Garrigues, M. Abou Hachem, K. Almdal, B. Svensson

269. Conjugating bile salts with modified dietary fiber polymers: A novel strategy to cholesterol lowering. J. Zornjak, Y. Dong, A. Esker, K.J. Edgar, **Cristina Fernandez Fraguas**

270. Synthetic arabinomannan glycolipids inhibit stationary phase survival of a mycobacteirum. **N. Jayaraman**, M.G. Krishna, S. Kirtimaan, D. Chatterji

271. Isolation, purification and characterization of a*phanothece sp.* exopolysaccharide as prebiotic candidate molecule. **K.S. Kiran**

272. Dual photo-affinity labeling probe for modification of lectin. **L. PeiJhen**, C. Lin

273. Engineering chitin deacetylases for the biotechnological production of chitosans with defined acetylation patterns. H. Aragunde, **X. Biarnés**, A. Planas

274. Quantitative analyses of NMR *J*-coupling ensembles in oligosaccharides: Glycosidic linkage conformational populations using circular statistics. W. Zhang, T. Turney, Q. Pan, M. Hadad, A. Coscia, **R. Meredith**, X. Wang, R.J. Woods, M. Yoon, I. Carmichael, A.S. Serianni

275. Application of click chemistry to access triazole-linked piperidine and azepane pseudo-disaccharide iminosugars with glycosidase inhibitory properties. L.B. Zamoner, V. Aragão-Leonetti, S.P. Mantoani, M.D. Rugen, S.A. Nepogodiev, R.A. Field, **I. Carvalho**

276. Label free detection of glycan-protein interactions for array development by surface enhanced Raman spectroscopy (SERS). **X. Li**, S.J. Martin, Z. Chinoy, L. Liu, R.A. Dluhy, G. Boons

277. Functional evaluation of membrane glycolipids using HaloTag system. **K. Kabayama**, S. Miyake, T. Yamaji, K. Hanada, Y. Manabe, K. Fukase

278. Development of oligosaccharide substrates and inhibitors for the *E. coli* de-*N*-acetylase PgaB. **A. Forman**, B. DiFrancesco, M. Nitz

279. Fully synthetic TACA-based anticancer vaccine prototypes. **C. Pifferi**, D. Goyard, B. Thomas, G. Daskhan, N. Berthet, O. Renaudet

280. A new method for the preparation of 2-acetyl-aryl-gluco- and galactopyranosides. **M. Nagornaya**

282. Exploring the mode of action of ipomoeassin natural glycoresins. **Z. Hu**, G. Zong, H. Aljewari, J. Zhou, Y. Du, W. Shi

283. Investigation of time-resolved localization and transformation using fluorescent glycosphingolipid probes. **H. Ohshika**, K. Arai, Y. Kanie, O. Kanie

284. Late-stage modification of *C-3* tigoyl moiety of ipomoeassin F to enable SAR studies. **L. Whisenhunt**, G. Zong, Z. Hu, H. Aljewari, W. Shi

285. Investigation of the conversion process of a fluorescently labeled lactosylsphingolipid in cultured CHO-K1 cells. Y. Kanie, M. Taniuchi, **O. Kanie**

286. Thio-functionalization of dihydrolevoglucosenone at C-3 *via* aldol condensation with thienyl aldehydes. **Z.J. Witczak**, R. Bielski

287. Novel *S*-glycosylated thiosemicarbazide act as potent anticancer agents in breast, cervix and ovarian cancer cells. A. Czubatka-Bienkowska, J. Sarnik, A. Macieja, **Z.J. Witczak**, T. Poplawski

288. Novel thio-sugar analog act as potent anticancer agent in cervical cancer cells through induction of oxidative stress. J. Sarnik, A. Czubatka-Bienkowska, A. Macieja, **Z.J. Witczak**, T. Poplawski

289. Examination of bond cleavage in transition state of *β*-glucosidase from *Thermotoga maritima*. **R. Zaenglein**, L.D. Byers

290. Glycan-conjugated fluorescent and magnetic nanoparticles to detect *Helicobacter* pylori. J. Hyun, **I. Shin**

TUESDAY EVENING, July 19

---

291. Carbohydrate microarrays for screening of functional glycans that stimulate the SIGN-R1 mediated immune response. **J. Hyun**, I. Shin

292. ^{13}C-Labeled idohexopyranosyl rings: Effects of methyl glycosidation and C6 oxidation on ring conformational equilibria. B. Bose-Basu, **W. Zhang**, J. Kennedy, M. Hadad, I. Carmichael, A.S. Serianni

293. Glycoside surfactants effective against *bacillus anthracis* and the mechanism of action. **J. Pais**, **A.S. Viana**, **A.P. Rauter**, **L. Sobral**, **R. Dias**, **R. Tenreiro**

294. Novel C-4 modified CD22 inhibitors as the basis for B-cell targeting nanoparticle. **P. Madge**, T. Haselhorst, A. Maggioni, S. Kelm, M. von Itzstein

295. Carbohydrate composition of cotton plant biomass materials. **Z. He**, D. Olk, H. Tewolde, M. Shankle, H. Zhang

296. Measurement of redundant NMR spin-spin couplings in ^{13}C-labeled high mannose oligosaccharides. **M. Yoon**, W. Zhang, A.S. Serianni

297. Characterization of the therapeutically active immune complex of anti-beta glucan antibodies and Imprime PGG, an innate immune activating carbohydrate PAMP. **M.E. Danielson**, K. Michel, A. Rollings, T. Jevne, P. Will, N. Bose, A. Bykowski Jonas, B. Harrison, S. Leonardo, A.S. Magee

298. Solid-state NMR measurements of ^{13}C-^{13}C spin-couplings in saccharides: Validation of the effects of conformation on coupling magnitudes. **W. Zhang**, J. Zajicek, **M. Yoon**, M. Hadad, M.H. Frey, A.G. Oliver, A.S. Serianni

299. Isolation and characterization of xylans cell wall from *Argania spinosa* leaves. **K. Hachem**, C. Faugeron, M. Kaid-Harche, V. Sol, V. Gloaguen

300. Optimized production of glycolipids using saccharide primer and animal cells cultured with fluorous solvent. **M.Z. Kasuya**, K. Hatanaka

301. Development of the novel tumor immunotherapy using α-gal epitope as an adjuvant. **K. Fukase**, H. Li, Y. Manabe, K. Tokunaga, N. Terao, S. Takamatsu, M. Tanemura, E. Miyoshi

302. Purification and characterization of α-1,3-glucosidase from *Aspergillus niger*. **M. Ma**, T. Tagami, M. Okuyama, A. Kimura

303. The α-galactosidase A-deficient rat: Characterization of glycosphingolipid expression in a new animal model of Fabry disease. **K. Aoki**, J.J. Miller, M. Tiemeyer, N. Dahms

TUESDAY EVENING, July 19

304. Ionic liquid tag as a multiple functional tag for electrochemical glycosylation. **N. Sasaki**, Y. Isoda, T. Nokami, T. Itoh

305. Specific N-glycans regulate cell surface expression of AMPA-type glutamate receptors. **J. Morise**, S. Yamamoto, N. Hashii, N. Kawasaki, H. Takematsu, S. Oka

306. *Trypanosoma cruzi* trans-sialidase: A tool for multisialylation of β-thio-glycoclusters. **R. Agusti**, M.E. Cano, A.J. Cagnoni, J. Kovensky, R.M. Lederkremer, M.L. Uhrig

307. A one-step Selective Exo-Enzymatic Labeling (SEEL) strategy for the biotinylation and identification of glycoproteins of living cells. **T. Sun**, S. Yu, P. Zhao, L. Meng, K.W. Moremen, L. Wells, R. Steet, G. Boons

308. Glucosyl glycoclusters with potential *α*-glucosidase inhibitory activities. **A. Figueredo**, I. Carvalho

309. Exploring contributions of shape and charge to inhibition of family GH99 glycosidases. **P.Z. Fernandes**, L. Sobala, G. Davies, S.J. Williams

310. Clickable saccharides for targeting dendritic cells. **J. Langhanki**, T. Opatz

311. Preparation of chito-oligosaccharide-dihydrazide hybrids. **I.V. Mo**, M.Ø. Dalheim, C. Schatz, B. Christensen

312. RG-I regions from elderflower pectins substituted on GalA are strong immunomodulators. **G. Ho**, Y. Zou, H. Wangensteen, H. Barsett

313. Competitive AlphaScreen® assay for hyaluronan detection. **X. Huang**, M.K. Cowman

314. Simple, direct conjugation of bacterial O-PS–core antigens to proteins: A closer look. **P. Xu**

315. Identifying the fractionated *N*-glycome of urothelial cells. **C. Wang**, R. Duke, P. Fisher, D. Ungar, J. Southgate, J. Thomas-Oates

316. Functional isomaltomegalosaccharide synthesized by enzymatic decoration. **Y. Kumagai**, W. Lang, J. Sadahiro, M. Okuyama, H. Hara, A. Kimura

317. Structure of brea gum (an exudate from *Cercidium australe*). **C. Stortz**, F. Sznaider, D.A. Navarro

318. A full conformational study of both anomers of D-galactopyranose by vacuum and solvent-simulated DFT calculations. **C. Stortz**, E. Del Vigo, C. Marino

TUESDAY EVENING, July 19

---

319. Quantification of dermatan sulfate from urine in Ehlers-Danlos syndrome caused by mutations in dermatan 4-*O*-sulfotransferae-1. **S. Mizumoto**, T. Kosho, A. Hatamochi, T. Honda, S. Yamada, K. Sugahara

320. Conformational preferences of acyclic aldo- and ketohexoses and their implication for anomeric equilibria. **W. Plazinski**, A. Plazinska, M. Drach

321. The interrupted Pummerer reaction in glycosylation. X. Xiao, X. Zhao, P. Shu, Y. Zhao, Y. Liu, J. Sun, J. Zeng, **Q. Wan**

322. Alginate - β-lactoglobulin particle formation as affected by alginate degree of polymerization. **E.P. Stender**, S. Khan, R. Ipsen, F. Madsen, P. Hägglund, M. Abou Hachem, K. Almdal, P. Westh, B. Svensson

323. Glucosylation of catechol with the GTFA glucansucrase enzyme from *Lactobacillus reuteri* and sucrose as donor substrate. **S.S. van Leeuwen**, E.M. Te Poele, P. Grijpstra, L. Dijkhuizen

324. Comparative structural characterization of 7 commercial galacto-oligosaccharide (GOS) products. **S.S. van Leeuwen**, B.J. Kuipers, J.P. Kamerling, L. Dijkhuizen

325. Glycolipids for photoswitching of molecular monolayer membranes. **F. Reise**, B.M. Murphy, T.K. Lindhorst

326. Enhancement of quercetin water solubility with steviol glucosides and the studies of biological properties. T. Nguyen, S. Yu, B. Gu, H. Lim, K. Hwang, J. Park, **D. Kim**

327. Baylis-Hillman reaction of HMF and GMF in water and biobased solvents: An atom economy access to new biobased intermediates. **Y. Queneau**, J. Tan, M. Ahmar

328. Catalytic aerobic oxidation of reducing sugars issued from softwood hemicellulose acid hydrolysis. **Y. Queneau**, M. Ahmar, E. Derrien, E. Martin-Sisteron, G. Raffin, C. Pinel, M. Besson, P. Marion

329. Self-assembling carbohydrate-functionalized graphene for capturing and killing *E. coli* bacteria. **P.M. Bharate**, Z. Qi, C. Lai, B. Ziem, C. Böttcher, A. Schulz, F. Beckert, B. Hatting, R. Mülhaupt, R. Haag, P.H. Seeberger

330. Anticoagulant properties of the rhamnan sulfate isolated from commercially cultured green seaweed, *Monostroma nitidum*. **M. Tako**

TUESDAY EVENING, July 19

331. Synthetic study on pseudaminic acid from pathogenic bacteria. **H. Tamai**, D.B. Werz

332. Molar mass distribution of O-antigens from *E. coli* strains O44, O125, O142 and O173. **G. Widmalm**, **B. Christensen**

333. A new strategy for the preparation of 2-acetyl aryl glycosides. **E. Stepanova**

334. Examination of active sites of *β*-galactosidases by interactions with galactonoamidine inhibitors. **J.B. Pickens**, S. Striegler

335. Automated glycan assembly of xylan oligosaccharides for probing plant cell wall glycan-directed antibodies. **D. Schmidt**, G. de Kruijff, F. Schuhmacher, P.H. Seeberger, F. Pfrengle

336. Automated glycan assembly of plant cell wall-derived arabinogalactan oligosaccharides for epitope mapping of cell wall glycan-directed antibodies. **M. Bartetzko**, F. Schuhmacher, P.H. Seeberger, F. Pfrengle

337. Protection against experimental melioidosis with a synthetic beta-manno-heptopyranose hexasaccharide glycoconjugate. A.E. Scott, W.J. Christ, A.J. George, M.G. Stokes, G.J. Lohman, Y. Guo, M. Jones, R. Titball, T.P. Atkins, **S. Campbell**, J.L. Prior

338. Diethylaminoethyl sepharose (DEAE-Sepharose) microcolumn for glycopeptides enrichment. **H. Zhu**, X. Li, J. Qu, K. Jiang, E. Gashash, C. Ma, P.G. Wang

339. On chip development of N-glycan mimetics for improving CLRs targeting. **A. Cioce**, S. Serna, A. Hernandez, G. Goti, A. Bernardi, N.C. Reichardt

340. High xylan content biodegradable films for food packaging. **M.S. Peresin**, C.N. Schnell, M.V. Galván, P. Mocchiutti, M.Á. Zanuttini

341. Understanding the blood-compatibility of cellulosic hemodialysis membranes. **M. Roman**, X. Tan

342. Hybrid molecular nanomaterials for carbohydrate-lectin recognition and self-assembly. M.E. Pérez-Ojeda, B. Trastoy, I. García-Moreno, **J.L. Chiara**

343. Effect of natural based plasticizers on physical properties of biodegradable starch-based films. **Y. Miao**, J. Hasjim, B. Pora

443. The synthesis of N-acetyllactosamine functionalized dendrimers and their role in galectin-3 mediated cell aggregation studies. **J. H. Ennist**, M. J. Cloninger

WEDNESDAY MORNING, July 20

PLENARY SESSION
Plenary Lecture (6)

New Orleans Marriott, Acadia Room
S. M. Murphy, *Organizer*; M. Pratt, *Presiding*

8:30 344. Accessing new and improved CAZymes through metagenomics, synthesis and directed evolution. **Stephen G. Withers**

HUDSON AWARD SYMPOSIUM IN HONOR OF GEERT-JAN BOONS
Claude S. Hudson Award Symposium
Cosponsored by the Carbohydrate Division of ACS

New Orleans Marriott, Acadia Room
J. C. Paulson, *Organizer, Presiding*; L. Wang, *Presiding*

9:15 345. An integrated approach to uncover ligands for heparan sulfate binding proteins. **Geert-Jan Boons**

10:00 Intermission, Exhibits, Posters – Preservation Hall

10:30 346. Keynote Lecture: Lectins as microbial detectors. **Laura L. Kiessling**

11:00 347. From stereocontrolled glycosylation to the automated oligosaccharide synthesis. **Alexei Demchenko**, S. Pistorio, K.J. Stine, J. Yasomanee

11:20 348. NMR investigations on the recognition of glycans by protein receptors. **Jesus Jimenez-Barbero**, A. Arda, A. Poveda, A. Gimeno, P. Valverde, L. Unione, H. Coelho, S. Delgado, A. Canales, F.J. Canada

11:40 349. Synthesis of heparan sulfate oligosaccharides and glycopeptides. **Xuefei Huang**, W. Yang, S.B. Dulaney, **Y. Xu**, P. Wang, K. Yoshida, B. Yang, J. Liu

12:00 350. Catalytic stereoselective synthesis of deoxyglycosides. **M. Carmen Galan**

12:20 351. Getting a grip on glycan readers in human-bacteria interactions. **Tom Wennekes**

WEDNESDAY MORNING, July 20

GLYCOBIOCHEMISTRY & GLYCOBIOLOGY
Structure Analysis Bioinformatics

New Orleans Marriott, Bissonet
L. L. Kiessling, J. C. Paulson, *Organizers*; A. P. Rauter, *Presiding*

10:30 352. Keynote Lecture: Functional glycomics: Our sugar-lined blood vessels. **Gene Cioffi**

11:00 353. Pentasaccharide: A novel oligosaccharide isolated from *Ovis aries* milk. **Ashok K. Ranjan**, D. Deepak

11:20 354. Regional differences in hMOS composition within Vietnam. **Sander S. van Leeuwen**, E. Stoutjesdijk, E.J. van Leusen - van Kan, A. Schaafsma, J.P. Kamerling, F.A. Muskiet, J. Dijck - Brouwer, L. Dijkhuizen

11:40 355. Selective inhibition of aggregation and toxicity of a Tau-derived peptide using its glycosylated analogs. **Daniel Segal**, M. Frenkel-Pinter, M. Richman, A. Belostozky, A. Abu-Mokh, E. Gazit, S. Rahimipour

12:00 356. Novel aqueous methylation chemistry and partial acid hydrolysis of switchgrass. **Michael A. Madson**

12:20 357. A tool box for identification and characterization of surface binding sites (SBSs) in carbohydrate-active enzymes. D. Cockburn, C. Wilkens, **Birte Svensson**

NEW DEVELOPMENTS IN GLYCOSCIENCE
Carbohydrates in Health and Disease (2)

New Orleans Marriott, Galerie 2
A. D. French, M. A. Godshall, S. M. Murphy, J. R. Vercellotti, *Organizers*; Y. A. Knirel, *Presiding*

10:30 358. The presence and structure of immunogenic form of enterobacterial common antigen (ECA_{LPS}) among rough *E. coli* R1, R2, R3 and R4 strains. **Marta Kaszowska**, A. Maciejewska, C. Lugowski, J. Lukasiewicz

11:00 359. Fluorescence light up probes for tumor specific cell surface glycan motifs. **Sudhirkumar A. Shinde**, K. Ciurkot, Z. El-Schich, A. Gjörloff Wingren, B. Sellergren

11:20 360. Eliciting fractions from parasitic fungi *Leptosphaeria maculans* and *Alternaria brassicicola*: Isolation and structural characterization. **Roman Bleha**, A. Synytsya, B. Jindrichova, L. Burketova, O. Valentova

11:40 361. Identification and structural analysis of D-galactan-III in lipopolysaccharides from *Klebsiella pneumoniae* O1 and O2. **Katarina Stojkovic**, M. Kaszowska, J. Lukasiewicz

12:00 362. N-glycoprotein biomarker identification in rheumatoid arthritis patients' sera. **Xu Li**, P.G. Wang, C. Ma, J. Li, H. Zhu, X. Li

12:20 363. A binding cooperativity study of lectin-mixed gangliosides on cell mimicking surface. **Hung-Jen Wu**, N. Worstell, P. Krishnan, J. Weatherston, A. Singla, C. Lee

WEDNESDAY MORNING, July 20

SUSTAINABLE SOLUTIONS FOR PROCESSES & PRODUCTS AT DUPONT

New Orleans Marriott, Galeria 3
R. DiCosimo, *Organizer, Presiding*

10:30 364. Keynote Lecture: Structure/function relationships in plant cellulose synthesis. **Candace H. Haigler**

11:00 365. Insights in enzyme development for the fuel ethanol industry by analysis and characterization of 'left-over' carbohydrates. **Pauline Teunissen**

11:20 366. Novel polysaccharides for industrial and consumer product applications. **Helen Lu**, C.P. Lenges, N. Behabtu, S.M. Hennessey, S.H. Privette, J. Fischer

11:40 367. Food hydrocolloids: Structure-function relationships. **Kenneth A. Botkjaer**

12:00 368. Efficient enzymatic synthesis of inulooligosaccharides. **Slavko Kralj**

12:20 369. Enzymatic production of soluble carbohydrate fiber for food ingredient applications. **Robert DiCosimo**, Q. Cheng, S. Rothman, Z. You, M. Payne

---

SYNTHESIS OF CARBOHYDRATES, GLYCOCONJUGATES & GLYCAN-BASED BIOMATERIALS
Synthetic Glycotherapeutics

New Orleans Marriott, Galerie 5, 6
G. Boons, M. Cloninger, A. Demchenko, *Organizers*; D. B. Werz, *Presiding*

10:30 370. Keynote Lecture: Supramolecular polysaccharide assemblies: A novel therapeutic strategy against mucosal microbial infections. **Kawthar Bouchemal**

11:00 371. Synthesis and evaluation of a small molecule carbohydrate with immunomodualtory properties. **Cecilia E. Marzabadi**

11:20 54. Stereoselective synthesis of oligo-L-arabinofuranosides, its derivatives and L-arabinofuranosylated probes for β-L-arabinofuranosidases. **Akihiro Ishiwata**, S. Kaeothip, K. Kimura, S. Narita, Y. Ito

11:40 373. Single-entity heparan sulfate glycomimetic clusters for therapeutic applications. **OlgaV. Zubkova**, P.C. Tyler, J.E. Turnbull

12:00 374. The design and synthesis of anti-virulence *C*-mannosides: A new approach to curing urinary tract infections. **L. McGrane**, Z. Han, Z. Cusumano, V. Kalas, J. Pinkner, S. Hultgren, J. Janetka

12:20 375. Modular synthetic approach to *C*-glycosides, *C*-disaccharides and *C*-glycoclusters. J. Choutka, L. Erhartová, R. Pohl, **K. Parkan**

THURSDAY MORNING, July 21

PLENARY SESSION
Plenary Lecture (7)

New Orleans Marriott, Acadia Room
S. M. Murphy, *Organizer* ; S. J. Williams, *Presiding*

8:30 376. A journey into the world of eukaryotic cell walls: Structure and biosynthesis of essential polysaccharides. **Vincent Bulone**

PLENARY SESSION
Plenary Lecture (8)

New Orleans Marriott, Acadia Room
S. M. Murphy, *Organizer*; B. Svensson, *Presiding*

9:15 377. Carbohydrate quality and how the concept may relate to healthier foods. **Bruce Hamaker**

10:00 Intermission, Exhibits, Posters – Preservation Hall

THURSDAY MORNING, July 21

DEREK HORTON MEMORIAL SYMPOSIUM

Financially supported by V-LABS, INC, Covington, Louisiana

New Orleans Marriott, Bissonet
W. Priebe, Z. J. Witczak, *Organizers, Presiding*

10:30 378. Special Lecture: The challenging role of carbohydrates at the interface of chemistry and biology. **Stephen Hanessian**

11:00 379. Derek Horton: Prolific scientist, editor, and wise mentor. **Zbigniew J. Witczak**

11:20 380. Derek Horton - the inspiration and the challenge. **Amelia P. Rauter**

11:40 381. Derek Horton as an amalgam of cultural assets and scientific historical opportunity. **John R. Vercellotti**

12:00 382. Exploring carbohydrate metabolism to target cancer: Derek Horton's inspiration. **Waldemar Priebe**

GLYCOTECHNOLOGY IN MEDICINE, NUTRITION & MATERIAL SCIENCE
Glycotechnology in Material Science

New Orleans Marriott, Galerie 3
M. A. Godshall, S. M. Murphy, *Organizers*; M. Roman, *Organizer, Presiding*

10:30 383. Keynote Lecture: Interfacial behavior of carbohydrates: from amphiphilic sugars to cellulose nanocrystals. **Orlando J. Rojas**

11:00 384. Smart microarray platforms for understanding biochemical interactions. **Caroline I. Biggs**, M.I. Gibson

11:20 385. Protecting-group-free synthesis of glycopolymers bearing sialyloligosaccharides via direct anomeric activation of free saccharides. **Tomonari Tanaka**, Y. Zhou, T. Takahashi, T. Suzuki, Y. Miura

11:40 386. Peptide-substituted alginate biomaterials. M.Ø. Dalheim, B.L. Strand, J. Vanacker, F.L. Aachmann, **Bjoern Christensen**

12:00-106. Design, synthesis, and activity of nanocellulosic protease sensors. V. Edwards, **Krystal R. Fontenot**, N.T. Prevost, A.D. French, B.D. Condon, F. Liebner, D. Haldane

THURSDAY MORNING, July 21

NEW DEVELOPMENTS IN GLYCOSCIENCE
General Carbohydrate Studies (1)

New Orleans Marriott, GaleriE 2
A. D. French, M. A. Godshall, S. M. Murphy, J. R. Vercellotti, *Organizers*; P. Kosma, *Presiding*

10:30 389. Genetic mechanisms of the O-antigen structure diversification of *Escherichia coli* and *Shigella* spp. **Yuriy A. Knirel**, S.N. Senchenkova, A.V. Perepelov, E.L. Zdorovenko, O.I. Naumenko, A.S. Shashkov, X. Guo, B. Liu, L. Wang

11:00 390. Maillard reaction inhibits non-enzymatic transglycosylation reactions: Relevance for melanoidins formation. A. Moreira, E. Maciel, P. Domingues, F. Nunes, M.M. Domingues, **Manuel A. Coimbra**

11:20 391. Progress towards a Streptococcus pneumoniae semi-synthetic glycoconjugate vaccine. **Petra Menova**, A. Calow, S.G. Parameswarappa, N. Khan, C.L. Pereira, P.H. Seeberger

11:40 392. Targeting tumor associated carbohydrate antigens (TACAs) to diagnose and stage cancer using synthetic lectin (SL) arrays. **John J. Lavigne**

12:00 157. Reversible glycan self assembled monolayers (rSAMs) for ultrasensitive virus sensing via enhanced multivalent interactions. **Sing Yee Yeung**, A. Tennboll, T. Arnebrant, B. Sellergren

SYNTHESIS OF CARBOHYDRATES, GLYCOCONJUGATES & GLYCAN-BASED BIOMATERIALS
Chemical Glycosylation

New Orleans Marriott, Acadia Room
G. Boons, M. Cloninger, A. Demchenko, *Organizers*; P. V. Murphy, *Presiding*

10:30 393. Keynote Lecture: Quick assembly of glycosides and glycoproteins: General tactics and new glycosylation methods. H. Chai, K. Hoang, **Xue-Wei Liu**

11:00 394. New methods for the direct synthesis of glycoconjugates from reducing sugars in water. **Antony J. Fairbanks**

11:20 395. Multifunctional glycocalyx-mimetic surfaces reduce platelet activation on titanium. **Matt Kipper**, K.C. Popat, M.M. Reynolds

11:40 396. Development of silyl-assisted 1,2-*cis*-α-glucosylation reactions. **Kichiro Totani**, Y. Shinoda, M. Shiba, S. Iwamoto, A. Koizumi, Y. Matsuzaki, M. Hirano

12:00 397. Templated oligosaccharide synthesis. **Xiao Jia**, A. Demchenko, P. Pornsuriyasak

THURSDAY MORNING, July 21

SYNTHESIS OF CARBOHYDRATES, GLYCOCONJUGATES & GLYCAN-BASED BIOMATERIALS
Synthesis of Plant Oligosaccharides

New Orleans Marriott, Galerie 5, 6
G. Boons, M. Cloninger, A. Demchenko, *Organizers*; D. Crich, *Presiding*

10:30 398. Keynote Lecture: Cellulose in water and other solvents - aligning and self-assembling. **S.J. Eichhorn**

11:00 399. Synthesis and application of oligosaccharides related to plant cell wall glycans. **M. Clausen**

11:20 400. Synthetic plant carbohydrates as a toolbox for cell wall biology. P. Dallabernardina, D. Schmidt, M.P. Bartetzko, C. Ruprecht, F. Schuhmacher, P.H. Seeberger, **F. Pfrengle**

11:40 401. Genetic control of starch biosynthesis in plants. **H. Cao**

12:00 402. An efficient synthetic route to bicyclic sugars: Tuning regio- and stereoselectivity on Wittig olefination by solvent and steric effects. **R. Nunes**, V. Cachatra, D. Vila-Viçosa, M. Calhorda, A.P. Rauter

THURSDAY AFTERNOON, July 21

GLYCOBIOCHEMISTRY & GLYCOBIOLOGY

New Orleans Marriott, Bissonet
L. L. Kiessling, J. C. Paulson, *Organizers*; X. Liu, *Presiding*

2:20 403. Chitin deacetylases: How do they define the de-*N*-acetylation pattern? H. Aragunde, X. Biarnés, **Antoni Planas**

2:40 404. Probing glycogen biosynthesis via a palladium-mediated enzyme switch. **Matthew K. Bilyard**, W.W. Yue, B.G. Davis

3:00 405. Characterization of potent *endo*-xyloglucanases illuminates the first committed step in xyloglucan utilization by the model soil saprophyte *Cellvibrio japonicus*. **Mohamed Attia**, J. Stepper, C. Nelson, J. Gardner, G. Davies, H. Brumer

3:20 406. A key pyranose-2-phosphate motif is responsible for both antibiotic import and quorum-sensing regulation in *Agrobacterium tumefaciens*. **Yves Queneau**, A. El Sahili, S. Li, J. Lang, C. Virus, S. Planamente, M. Ahmar, B.G. Guimaraes, M. Aumont-Nicaise, A. Vigouroux, L. Soulère, J. Reader, D. Faure, S. Morera

3:40 Intermission – Preservation Hall

THURSDAY AFTERNOON, July 21

GLYCOTECHNOLOGY IN MEDICINE, NUTRITION & MATERIAL SCIENCE
Glycotechnology in Medicine

New Orleans Marriott, Galerie 3
M. A. Godshall, S. M. Murphy, M. Roman, *Organizers*; K. Bouchemal, *Presiding*

2:20 407. A synthetic TTox-glycopeptide vaccine against human, tumor-associated MUC-1 enabled the generation of a monoclonal antibody suitable for clinical applications. **Natascha Stergiou**, B. Palitzsch, M. Glaffig, J. Nagel, H. Kunz, E. Schmitt

2:40 408. Carbohydrate-based imaging probes for targeted alpha-particle therapy of metastatic melanoma. **Izabela Tworowska**, N. Wagh, M. Li, K. Ershadi, M.K. Schultz

3:00 409. Synthesis and biomedical applications of parasite glycans. K. Brzezicka, S. Serna, **Niels C. Reichardt**

3:20 410. Molecular basis of the interaction between the Group B Streptococcus Type III capsular polysaccharide and a protective monoclonal antibody. **Roberto Adamo**, F. Carboni, M. Fabbrini, R. De Ricco, V. Cattaneo, B. Brogioni, D. Veggi, V. Pinto, I. Passalacqua, D. Oldrini, D. Maione, E. Malito, I. Margarit, F. Berti

3:40 Intermission – Preservation Hall

NEW DEVELOPMENTS IN GLYCOSCIENCE
Structure and Analysis

New Orleans Marriott, Galerie 2
A. D. French, M. A. Godshall, S. M. Murphy, J. R. Vercellotti, *Organizers*; C. H. Haigler, *Presiding*

2:20 411. Hierarchical architecture of cellulose and its interaction mechanism with different plant cell wall polysaccharides investigated by small angle scattering techniques. **Marta Martinez Sanz**, P. Lopez-Sanchez, D. Mikkelsen, B. Flanagan, M. Gidley, E. Gilbert

2:40 412. NMR studies of compositional heterogeneity in polysaccharides. **H.N. Cheng**

3:00 413. Protein bioseparation using magnetically responsive cellulose nanocrystals. J. Guo, **Erkko Filpponen**, O.J. Rojas

3:20 414. Systematic quantification of cell surface glycoproteins by mass spectrometry. **Ronghhu Wu**

3:40 Intermission – Preservation Hall

THURSDAY AFTERNOON, July 21

SYNTHESIS OF CARBOHYDRATES, GLYCOCONJUGATES & GLYCAN-BASED BIOMATERIALS
Novel Synthetic Methods (1)

New Orleans Marriott, Acadia Room
G. Boons, M. Cloninger, A. Demchenko, *Organizers*; M. Galan, *Presiding*

2:20 415. A general chemoenzymatic strategy for the synthesis of glycosphingolipids. **Yunpeng Liu**, W. Liuqing, L. Li, H. Yu, X. Chen, P.G. Wang, J. Fang

2:40 416. Visible-light-promoted *O*-glycosylation with selenoglycoside and thioglycoside donors. **Justin R. Ragains**

3:00 417. Studies directed towards the synthesis of branched-backbone oligosaccharides of the pectic RG-I plant cell wall polysaccharide. **Shahid I. Awan**, M. Clausen

3:20 418. New principles of stereoselective carbohydrate synthesis. **Maciej A. Walczak**

3:40 Intermission – Preservation Hall

SYNTHESIS OF CARBOHYDRATES, GLYCOCONJUGATES & GLYCAN-BASED BIOMATERIALS
Synthesis of N-Glycans

New Orleans Marriott, Galerie 5, 6
G. Boons, M. Cloninger, A. Demchenko, *Organizers*; T. L. Lowary, *Presiding*

2:20 419. Synthesis of *N*-glycosyl compounds of D-glucuronamide derivatives as nucleoside analogs or as prospective mimics of glycosyl phosphates or nucleotides. **Nuno Manuel Xavier**, A. Porcheron, V.J. Martins, A. Fortuna, D. Batista, R. Jorda

2:40 420. Cascade reactions for the synthesis of modified *N*-acetyllactosamine oligomers and their multivalent presentation for specific galectin recognition. **Dominic Laaf**, S. Böcker, H. Pelantová, V. Kren, L. Elling

3:00 421. Synthesis of GlcA, GalNAc4S, and GalNAc3S6S derivatives as an inhibitor against Japanese encephalitis virus. **Miho Sakuragi**, A. Suzuki, M. Noto, S. Yasuda, K. Kato, R. Kawamura, K. Hidari, H. Nakano

3:20 422. Chemical synthesis of core fucosylated *N*-glycan: Precursor to eukaryotic glycan mimics. **Apoorva Srivastava**, I.A. Gagarinov, R. de Vries, G. Boons

3:40 Intermission – Preservation Hall

THURSDAY AFTERNOON, July 21

GLYCOBIOCHEMISTRY & GLYCOBIOLOGY

New Orleans Marriott, Bissonet
L. L. Kiessling, J. C. Paulson, *Organizers*; J. R. Vercellotti, *Presiding*

4:00 423. Discovery of a dedicated sulfoquinovosidase: A gateway enzyme to sulfoglycolysis. E. Goddard-Borger, G. Speciale, Y. Jin, G. Davies, **Spencer J. Williams**

4:20 424. Inhibitor-based manipulation reveals specific glycosignal production through dual mannose-trimming pathways in ER glycoprotein quality control. **Taiki Kuribara**, M. Hirano, G. Speciale, S.J. Williams, Y. Ito, K. Totani

4:40 425. Iminosugars as pharmacological chaperones for the treatment of LSDs. **Farah Lamiable-Oulaidi**, S. Front, E. Gallienne, P. Compain, O.R. Martin

5:00 426. Unusual active site structure of glycoside hydrolase family 116 and mechanistic implications. **James R. Cairns**, S. Pengthasong, R. Charoenwattanasatien, G. Kurisu

---

GLYCOTECHNOLOGY IN MEDICINE, NUTRITION & MATERIAL SCIENCE
Influence of Structure on Performance of Polysaccharides

New Orleans Marriott, Galeria 3
M. A. Godshall, S. M. Murphy, M. Roman, *Organizers*; J. K. Frost, *Presiding*

4:00 427. Evidence for supramolecular binding between cereal β-glucan and nutritionally relevant ligands. **Laura Nystroem**, S. Boulos

4:20 428. Brewing modifies yeast cell wall polysaccharides generating by-products to be exploited in diversified formulations and materials. **Elisabete Coelho**, S. Messias, R. Bastos, E. Correia, M. Pinto, M.A. Coimbra

4:40 429. The potential of marine polysaccharides for biomedical applications: The influence of structure on biological properties. **Claudia D. Nunes**, A. Rocha, A.S. Ferreira, C. Oliveira, T.H. Silva, A. Martins, N.M. Neves, R. Reis, M. Vilanova, M.A. Coimbra

5:00 430. Enzyme engineering enables an efficient enzymatic production processes for human milk oligosaccharides. **Sebastian Bartsch**, A. Vogel, R. Schmiedel, E. Champion, G. Dekany

THURSDAY AFTERNOON, July 21

NEW DEVELOPMENTS IN GLYCOSCIENCE
General Carbohydrate Studies (2)

New Orleans Marriott, Galeria 2
A. D. French, M. A. Godshall, S. M. Murphy, J. R. Vercellotti, *Organizers*; A. M. Striegel, *Presiding*

4:00 431. Probing the structure of resistant dextrins using ion-trap mass spectrometry. V. Quintano Ramos, **Sean Austin**

4:20 432. Metabolic labelling of cholesteryl glucosides in *Helicobacter pylori* reveals how uptake of human lipids enhances bacterial virulence. **Chun-Hung Lin**

4:40 433. Effect of a single point mutation on the interaction of glucans with a glucansucrase from *Leuconostoc mesenteroides* NRRL B-1118. **Gregory L. Coté**

5:00 434. Differential agglutination from *Dioclea grandiflora* lectins on red blood cells. **Tania Cortázar** , E.A. Reyes, N.A. Vega

---

SYNTHESIS OF CARBOHYDRATES, GLYCOCONJUGATES & GLYCAN-BASED BIOMATERIALS
Chemoenzymatic Synthesis

New Orleans Marriott, Acadia Room
G. Boons, M. Cloninger, A. Demchenko, *Organizers*; R. Daniellou, *Presiding*

4:00 435. Solving the diversity of glycans via core synthesis/enzymatic extension. **Peng G. Wang**

4:20 436. Chemoenzymatic synthesis of HIV glycopeptide antigens for vaccine design. **Lai-Xi Wang**

4:40 437. In vitro production of sugar phosphates and activated sugars using thermostable enzymes. **Leila Aminova**

5:00 438. Chemoenzymatic synthesis of carbohydrates. **Xi Chen**

THURSDAY AFTERNOON, July 21

SYNTHESIS OF CARBOHYDRATES, GLYCOCONJUGATES & GLYCAN-BASED BIOMATERIALS
Novel Synthetic Methods (2)

New Orleans Marriott, Galerie 5, 6
G. Boons, M. Cloninger, A. Demchenko, *Organizers;* M. Fridman, *Presiding*

4:00 439. Simple access to L-*glycero*-D-*manno* heptose at scale – a milestone towards more convenient syntheses of bacterial LPS-substructures. **Christian Stanetty**, I.R. Baxendale, M.D. Mihovilovic

4:20 440. Development of fluorescent ganglioside analogs for imaging of raft-based interactions. **Naoko Komura**, K. Suzuki, H. Ando, M. Konishi, A. Imamura, H. Ishida, A. Kusumi, M. Kiso

4:40 441. Synthesis and immunomodulatory activities of natural and designed glycoconjugates for understanding of NKT-cell activation. T. Aiba, S. Inuki, N. Hirata, S. Suehara, E. Kashiwabara, S. Nakagawa, A. Shimoyama, K. Fukase, **Yukari Fujimoto**

5:00 442. Synthesis of α-dystroglycan-derived glycopeptide with diagnostic and therapeutic applications In muscular dystrophies and cancer. **Vanessa L. Leiria Campo**, M. Marchiori, L. Rodrigues, M. Dias-Baruffi

CLOSING CEREMONY

Closing Remarks, Poster Prizes, and Future Meetings

New Orleans Marriott, Acadia Room
Sheila M. Murphy, Organizer
Alfred D. French, *Presiding*

5:20 Closing Remarks. **Alfred D. French** (Symposium Chair)

5:30 Poster Recognition. **H.N. Cheng**

5:35 19th European Carbohydrate Symposium in Barcelona. **Antoni Planas**

5:45 29th International Carbohydrate Symposium in Lisbon. **Amelia Rauter**

5:55 Final Remarks. **Alfred D. French**

6:40. Buses begin leaving for the Closing Reception and Banquet at Mardi Gras World

Author and Paper Number List

International Carbohydrate Symposium (ICS)
1155 16th St NW
Washington, DC 20036
USA

ISBN 978-1-5108-3585-6
9 781510 835856

CURRAN ASSOCIATES INC.
proceedings.com

National Conference on the Role of Polish Coal in the National and European Energy Sector 2017

IOP Conference Series: Materials Science and Engineering
Volume 268

Andrychow, Poland
19 - 21 June 2017

IOP
Institute of Physics